STUDENT HANDBOOK OF FORMULAS, DATA AND EQUATIONS

NOGGLE

PHYSICAL CHEMISTRY

Third Edition

HarperCollins*College*Publishers

HarperCollins® and ▇® are registered trademarks of HarperCollins Publishers Inc.

Student Handbook of Formulas, Data and Equations to accompany Noggle, *Physical Chemistry*, Third Edition

ISBN 0-673-52342-X

96 97 98 99 00 9 8 7 6 5 4 3 2 1

QD453.2·N64

Periodic Table of the Elements

1 ← IUPAC Group Number

IA ← Traditional American Group Number

Period																		
	1 1 H 1.0079	**2** IIA											**13** IIIA	**14** IVA	**15** VA	**16** VIA	**17** VIIA	**18** 0 2 He 4.00260

1	1 H 1.0079	2 IIA											13 IIIA	14 IVA	15 VA	16 VIA	17 VIIA	2 He 4.00260
2	3 Li 6.941	4 Be 9.01218											5 B 10.81	6 C 12.011	7 N 14.0067	8 O 15.9994	9 F 18.9984	10 Ne 20.179
3	11 Na 22.9898	12 Mg 24.305	3 IIIB	4 IVB	5 VB	6 VIB	7 VIIB	8	9 VIIIB	10	11 IB	12 IIB	13 Al 26.9815	14 Si 28.0855	15 P 30.9738	16 S 32.06	17 Cl 35.453	18 Ar 39.948
4	19 K 39.0983	20 Ca 40.08	21 Sc 44.9559	22 Ti 47.88	23 V 50.9415	24 Cr 51.996	25 Mn 54.9380	26 Fe 55.847	27 Co 58.9332	28 Ni 58.69	29 Cu 63.546	30 Zn 65.39	31 Ga 69.72	32 Ge 72.59	33 As 74.9216	34 Se 78.96	35 Br 79.904	36 Kr 83.80
5	37 Rb 85.4678	38 Sr 87.62	39 Y 88.9059	40 Zr 91.224	41 Nb 92.9064	42 Mo 95.94	43 Tc (98)	44 Ru 101.07	45 Rh 102.906	46 Pd 106.42	47 Ag 107.868	48 Cd 112.41	49 In 114.82	50 Sn 118.71	51 Sb 121.75	52 Te 127.60	53 I 126.905	54 Xe 131.29
6	55 Cs 132.905	56 Ba 137.33	71 Lu 174.967	72 Hf 178.49	73 Ta 180.948	74 W 183.85	75 Re 186.207	76 Os 190.2	77 Ir 192.22	78 Pt 195.08	79 Au 196.967	80 Hg 200.59	81 Tl 204.383	82 Pb 207.2	83 Bi 208.980	84 Po (209)	85 At (210)	86 Rn (222)
7	87 Fr (223)	88 Ra 226.025	103 Lr (262)	104 Unq (261)	105 Unp (262)	106 Unh (266)	107 Uns (262)	108 Uno (265)	109 Une (266)									

57 La 138.906	58 Ce 140.12	59 Pr 140.908	60 Nd 144.24	61 Pm (145)	62 Sm 150.36	63 Eu 151.96	64 Gd 157.25	65 Tb 158.925	66 Dy 162.50	67 Ho 164.930	68 Er 167.26	69 Tm 168.934	70 Yb 173.04
89 Ac 227.028	90 Th 232.038	91 Pa 231.036	92 U 238.029	93 Np (237)	94 Pu (244)	95 Am (243)	96 Cm (247)	97 Bk (247)	98 Cf (251)	99 Es (252)	100 Fm (257)	101 Md (258)	102 No (259)

1 ← Atomic Number
H ← Element Symbol
1.0079 ← Relative Atomic Mass—values in parentheses are mass numbers for the most stable known isotope

Traditional Group Names:

1—alkali metals 2—alkaline earths 15—pnicogens 16—chalcogens 17—halogens 18—rare gases
3–12—transition metals La–Yb—lanthanides (rare earth metals) Ac–No—actinides Lr–Une—transactinides

IUPAC Commission on Nomenclature of Inorganic Chemistry Interim Provisional Names:
104—Unq (unnilquadium) 105—Unp (unnilpentium) 106—Unh (unnilhexium)
107—Uns (unnilseptium) 108—Unq (unniloctium) 109—Une (unnilennium)

IUPAC Commission on Nomenclature of Inorganic Chemistry Recommended Symbols and Names:
104—Db (dubnium) 105—Jl (joliotium) 106—Rf (rutherfordium)
107—Bh (bohrium) 108—Hn (hahnium) 109—Mt (meitnerium)

Discoverer's Recommended Symbols and Names:
104—Rf (rutherfordium) 105—Ha (hahnium) 106—Sg (seaborgium)
107—Ns (nielsbohrium) 108—Hs (hassium) 109—Mt (meitnerium)

Unit Conversions

Energy: $J = kg\ m^2\ s^{-2} = 10^7\ ergs;\ erg = g\ cm^2\ s^{-2}$

	J	cal	BTU	cm^3 atm
1 J	= 1	0.2390	9.488×10^{-4}	9.869
1 cal	= 4.184	1	3.96340×10^{-3}	41.293
1 BTU	= 1055.66	252.309	1	1.04186×10^4
1 cm^3 atm	= 0.101325	0.024217	9.59826×10^{-5}	1

Molecular Energy: rydberg = hartree/2

	cm^{-1}	eV	K	kJ/mol
1 cm^{-1}	= 1	1.23984×10^{-4}	1.438769	0.01196266
1 eV	= 8065.54	1	11604.5	96.4853
1 hartree	= 219474.63	27.2114	315773	2625.500

Pressure: $Pa = N\ m^{-2} = J\ m^{-3} = kg\ m^{-1}\ s^{-2} = 10\ dyne/cm^2$

	Pa	bar	atm	torr	psi
1 Pa	= 1	10^{-5}	9.86923×10^{-6}	7.5062×10^{-3}	1.4504×10^{-4}
1 bar	= 100000	1	0.986923	750.062	14.504
1 atm	= 101325	1.01325	1	760	14.696
1 torr	= 133.322	1.3332×10^{-3}	1.31579×10^{-3}	1	0.019337
1 psi	= 6894.7	0.0689487	0.068046	51.715	1

Force: $N = kg\ m\ s^{-2} = 10^5\ dyne;\ dyne = g\ cm\ s^{-2}$
Volume: $m^3 = 10\ liter = 10^6\ cm^3 = 35.315\ cu\ ft = 264.2\ gal\ (U.S.)$
Viscosity: (poise) $p = g\ cm^{-1}\ s^{-1}$; (SI) $Pa\ s = kg\ m^{-1}\ s^{-1} = 10\ p$
Temperature: $K = {}^\circ C + 273.15;\ {}^\circ R = {}^\circ F + 459.69 = 1.8\ K$
Length: $\text{Å} = 10^{-8}\ cm = 0.1\ nm = 100\ pm;\ inch = 2.54\ cm;\ mi = 1609.244\ m$
Miscellaneous English Units:

BTU/lb = 2.324 J/g   kg = 2.2046 lb
BTU/lb^{-1} $^\circ$R^{-1} = 4.184 J K^{-1} g^{-1} lb/cu ft = 16.02 kg m^{-3}

Constants

Name	Symbol	SI	cgs
Avogadro constant	L	6.0221367×10^{23} mol^{-1}	
Bohr magneton	μ_B	9.274015×10^{-24} J/T	$- \times 10^{-21}$ erg/gauss
Bohr radius	a_0	$5.2917725 \times 10^{-11}$ m	$- \times 10^{-9}$ cm
Boltzmann constant	k_b	1.380658×10^{-23} J/K	$- \times 10^{-16}$ erg/K
Debye	D	3.33564×10^{-30} C m	1 esu cm
Electron charge	e	$1.60217733 \times 10^{-19}$ C	4.80321×10^{-10} esu
Electron rest mass	m_e	$9.1093897 \times 10^{-31}$ kg	$- \times 10^{-28}$ g
Faraday constant	$\mathcal{F}$	96485 C mol^{-1}	
Gas constant	R	8.31451 J K^{-1} mol^{-1}	$- \times 10^{7}$ erg K^{-1} mol^{-1}
Gravity acceleration	g	9.80665 m s^{-2}	980.665 cm s^{-2}
Hartree energy	E_h	$4.3597482 \times 10^{-18}$ J	$- \times 10^{-11}$ erg
Light speed in vacuum	c	2.99792458×10^{8} m/s	$- \times 10^{10}$ cm/s
Permittivity of vacuum	ϵ_0	$8.8541878 \times 10^{-12}$ C^2 J^{-1} m^{-1}	$1/4\pi$ esu^2 erg^{-1} cm^{-1}
Planck's constant	h	$6.6260755 \times 10^{-34}$ J s	$- \times 10^{-27}$ erg s
Proton rest mass	m_p	1.672623×10^{-27} kg	$- \times 10^{-24}$ g

Miscellaneous: R = 1.9872 cal K^{-1} mol^{-1} = 82.056 cm^3 atm K^{-1} mol^{-1}; π = 3.1415926536; e = 2.718281828459; ln 10 = 2.302585093; hc/k = 1.4388 cm K; $RT/\mathcal{F}$ = 25.693 mV at 25°C.

Powers of 10: 10^x

x	Name	Symbol	x	Name	Symbol
-1	deci[a]	d	1	deca[a]	da
-2	centi[a]	c	2	hecto[a]	h
-3	milli	m	3	kilo	k
-6	micro	μ	6	mega	M
-9	nano	n	9	giga	G
-12	pico	p	12	tera	T
-15	femto	f	15	peta	P
-18	atto	a	18	exa	E

[a] To be avoided except for area and volume.

The Greek Alphabet

A, α	Alpha		N, ν	Nu
B, β	Beta		Ξ, ξ	Xi
Γ, γ	Gamma		O, o	Omicron
Δ, δ	Delta		Π, π	Pi
E, ϵ	Epsilon		P, ρ	Rho
Z, ζ	Zeta		Σ, σ	Sigma
H, η	Eta		T, τ	Tau
Θ, θ	Theta		Υ, υ	Upsilon
I, ι	Iota		Φ, ϕ	Phi
K, κ	Kappa		X, χ	Chi
Λ, λ	Lambda		Ψ, ψ	Psi
M, μ	Mu		Ω, ω	Omega

Table of Integrals

(Add a constant of integration to indefinite integrals.)

$$\int x^n dx = \frac{1}{n+1}x^{n+1} \quad (n \neq -1)$$

$$\int x^2 \sin^2 x\, dx = \frac{x^3}{6} - \left(\frac{x^2}{4} - \frac{1}{8}\right)\sin 2x - \frac{x\cos 2x}{4}$$

$$\int \frac{dx}{x} = \ln x$$

$$\int_0^\infty x^n e^{-ax} dx = \frac{n!}{a^{n+1}}$$

$$\int e^{ax} dx = \frac{e^{ax}}{a}$$

$$\int_0^\infty e^{-ax^2} dx = \frac{1}{2}\sqrt{\frac{\pi}{a}}$$

$$\int xe^{ax} dx = \frac{e^{ax}}{a^2}(ax - 1)$$

$$\int_0^\infty xe^{-ax^2} dx = \frac{1}{2a}$$

$$\int \frac{dx}{(a+bx)} = \frac{1}{b}\ln(a+bx)$$

$$\int_0^\infty x^2 e^{-ax^2} dx = \frac{1}{4a}\sqrt{\frac{\pi}{a}}$$

$$\int \frac{x\,dx}{(a+bx)} = \frac{1}{b^2}\{a+bx - a\ln(a+bx)\}$$

$$\int_0^\infty x^3 e^{-ax^2} dx = \frac{1}{2a^2}$$

$$\int \frac{dx}{x(a+bx)} = \frac{1}{a}\ln\left(\frac{x}{a+bx}\right)$$

$$\int_0^\infty x^4 e^{-ax^2} dx = \frac{3}{8a^2}\sqrt{\frac{\pi}{a}}$$

$$\int \sin x\, dx = -\cos x$$

$$\int_0^\infty \sqrt{x}\, e^{-ax} dx = \frac{1}{2a}\sqrt{\frac{\pi}{a}}$$

$$\int \cos x\, dx = \sin x$$

$$\int \sin^2 x\, dx = \frac{1}{2}x - \frac{1}{4}\sin 2x$$

$$\int \sin^3 x\, dx = -\frac{1}{3}\cos x\,(\sin^2 x + 2)$$

$$\int \cos^2 x\, dx = \frac{1}{2}x + \frac{1}{4}\sin 2x$$

$$\int \cos^3 x\, dx = \frac{1}{3}\sin x\,(\cos^2 x + 2)$$

$$\int \sin^2 x\,\cos x\, dx = \frac{1}{3}\sin^3 x$$

$$\int \sin nx \sin mx\, dx = \frac{\sin(n-m)x}{2(n-m)} - \frac{\sin(n+m)x}{2(n+m)}$$

$$\int \cos^2 x \sin x\, dx = -\frac{1}{3}\cos^3 x$$

$$\int x\sin^2 x\, dx = \frac{x^2}{4} - \frac{x\sin 2x}{4} - \frac{\cos 2x}{8}$$

Atomic Numbers and Atomic Weights

Element	Symbol	No.	Weight	Element	Symbol	No.	Weight
Actinium	Ac	89	227.0278	Molybdenum	Mo	42	95.94
Aluminum	Al	13	26.98154	Neodymium	Nd	60	144.24
Americium	Am	95	(243)	Neon	Ne	10	20.179
Antimony	Sb	51	121.75	Neptunium	Np	93	237.0482
Argon	Ar	18	39.948	Nickel	Ni	28	58.69
Arsenic	As	33	74.9216	Niobium	Nb	41	92.9064
Astatine	At	85	(210)	Nitrogen	N	7	14.0067
Barium	Ba	56	137.33	Nobelium	No	102	(259)
Berkelium	Bk	97	(247)	Osmium	Os	76	190.2
Beryllium	Be	4	9.01218	Oxygen	O	8	15.9994
Bismuth	Bi	83	208.9804	Palladium	Pd	46	106.42
Boron	B	5	10.811	Phosphorus	P	15	30.97376
Bromine	Br	35	79.904	Platinum	Pt	78	195.08±3
Cadmium	Cd	48	112.41	Plutonium	Pu	94	(244)
Calcium	Ca	20	40.078	Polonium	Po	84	(209)
Californium	Cf	98	(251)	Potassium	K	19	39.0983
Carbon	C	6	12.011	Praseodymium	Pr	59	140.9077
Cerium	Ce	58	140.12	Promethium	Pm	61	(145)
Cesium	Cs	55	132.9054	Protactinium	Pa	91	231.0359
Chlorine	Cl	17	35.453	Radium	Ra	88	226.0254
Chromium	Cr	24	51.9961	Radon	Rn	86	(222)
Cobalt	Co	27	58.9332	Rhenium	Re	75	186.207
Copper	Cu	29	63.546	Rhodium	Rh	45	102.9055
Curium	Cm	96	(247)	Rubidium	Rb	37	85.4678
Dysprosium	Dy	66	162.50	Ruthenium	Ru	44	101.07
Einsteinium	Es	99	(252)	Samarium	Sm	62	150.36
Erbium	Er	68	167.26	Scandium	Sc	21	44.95591E
uropium	Eu	63	151.96	Selenium	Se	34	78.96
Fermium	Fm	100	(257)	Silicon	Si	14	28.0855
Fluorine	F	9	18.998403	Silver	Ag	47	107.8682
Francium	Fr	87	(223)	Sodium	Na	11	22.98977G
adolinium	Gd	64	157.25	Strontium	Sr	38	87.62
Gallium	Ga	31	69.723	Sulfur	S	16	32.066
Germanium	Ge	32	72.59	Tantalum	Ta	73	180.9479
Gold	Au	79	196.9665	Technetium	Tc	43	(98)
Hafnium	Hf	72	178.49	Tellurium	Te	52	127.60
Helium	He	2	4.002602	Terbium	Tb	65	158.9254
Holmium	Ho	67	164.9304	Thallium	Tl	81	204.383
Hydrogen	H	1	1.00794	Thorium	Th	90	232.0381
Indium	In	49	114.82	Thulium	Tm	69	168.9342
Iodine	I	53	126.9045	Tin	Sn	50	118.710
Iridium	Ir	77	192.22	Titanium	Ti	22	47.88
Iron	Fe	26	55.847	Tungsten	W	74	183.85
Krypton	Kr	36	83.80	Unnilhexium	Unh	106	(263)
Lanthanum	La	57	138.9055	Unnilpentium	Unp	105	(262)
Lawrencium	Lr	103	(260)	Unnilquadium	Unq	104	(261)
Lead	Pb	82	207.2	Unnilseptium	Uns	107	(262)
Lithium	Li	3	6.941	Uranium	U	92	238.0289
Lutetium	Lu	71	174.967	Vanadium	V	23	50.9415
Magnesium	Mg	12	24.305	Xenon	Xe	54	131.29
Manganese	Mn	25	54.9380	Ytterbium	Yb	70	173.04
Mendelevium	Md	101	(258)	Yttrium	Y	39	88.9059
Mercury	Hg	80	200.59	Zinc	Zn	30	65.39
				Zirconium	Zr	40	91.224

A value in parentheses is the mass number of the isotope of longest half-life. Values in this table are from the IUPAC report "Atomic Weights of the Elements 1983," *Pure and Applied Chemistry,* Vol. 56, No. 6 (June 1984), pp. 653–674.

CHAPTER 1

Table 1.1   Gas Constants

	Critical Constants			Redlich-Kwong Constants		van der Waals Constants	
	T_c	P_c	$10^6 V_c$	a	$10^6 b$	a	$10^6 b$
	K	MPa	m³/mol	(see note)	m³/mol	Pa m⁶ mol⁻²	m³/mol
Ar	151.	4.955	75.2	1.671	21.95	0.1342	31.67
CH_4	190.6	4.641	98.8	3.194	29.59	0.2283	42.69
$CHCl_2F$	451.7	5.17	197	24.80	62.97	1.1515	90.85
Cl_2	417	7.711	123	13.61	38.96	0.6577	56.21
CO	134	3.546	90	1.732	27.22	0.1477	39.27
CO_2	304.2	7.397	95.6	6.448	29.63	0.3649	42.74
C_2H_2	309	6.282	60.1	7.895	35.43	0.4433	51.12
C_2H_4	282.9	5.157	127.5	7.713	39.51	0.4526	57.01
C_2H_6	305.33	4.871	136	9.882	45.15	0.5581	65.14
C_3H_6	364.6	4.62	181	16.23	56.84	0.8391	82.01
C_3H_8	369.9	4.256	200	18.27	62.61	0.9377	90.34
$C_4H_8{}^a$	419.6	4.023	240	26.5	75.14	1.2765	108.41
C_4H_{10}	425.1	3.797	232	29.00	80.66	1.3882	116.37
$C_5H_{12}{}^b$	469.6	3.37	311	41.91	100.4	1.908	144.8
$C_5H_{12}{}^c$	460.6	3.33	308	40.41	99.64	1.858	143.8
$C_5H_{12}{}^d$	433.8	3.20	303	36.20	97.66	1.715	140.9
HCl	325	8.268	86.2	6.806	28.32	0.3726	40.85
HBr	363	8.51	—	8.718	30.73	0.4516	44.33
HI	423	8.2	—	13.26	37.16	0.6364	53.61
He	5.3	0.229	57.7	0.008345	16.67	0.003577	24.05
H_2	33.2	1.297	65.0	0.1447	18.44	0.02479	26.60
Kr	210.6	5.492	92	3.464	27.62	0.2355	39.86
Ne	44.5	2.624	41.6	0.1488	12.22	0.022	17.62
NH_3	405.3	11.3	72.3	8.65	25.84	0.4241	37.28
N_2	126	3.394	90.0	1.551	26.74	0.1364	38.58
O_2	154.3	5.036	74.4	1.736	22.07	0.1379	31.84
SF_6	318.7	3.76	200	14.25	61.06	0.7878	88.09
SO_2	430	7.873	122	14.39	39.34	0.6849	56.76
SO_3	491.4	8.49	126	18.63	41.69	0.83	60.16
SiH_4	269.6	4.84	—	7.287	40.13	0.438	57.89
SiF_4	259	3.72	—	8.576	50.15	0.526	72.36
$SiCl_4$	506	3.76	—	45.27	96.94	1.986	139.9
Xe	290	5.88	119	7.18	35.5	0.42	51.21

Note: Unit for RK a constant is Pa m⁶ mol⁻² K$^{1/2}$.
[a] 1-butene
[b] *n*-pentane
[c] *iso*-pentane
[d] neopentane

van der Waals equation
$$P = \frac{RT}{V_m - b} - \frac{a}{V_m^2} \qquad (1.3b)$$

$$b = 2\pi L \frac{\sigma^3}{3} \qquad (1.4)$$

Redlich–Kwong (RK) equation

$$P = \frac{RT}{V_m - b} - \frac{a}{\sqrt{T}} \frac{1}{V_m(V_m + b)} \tag{1.5}$$

$$z \equiv \frac{PV_m}{RT} \tag{1.8}$$

$$z = 1 + \frac{B(T)}{V_m} + \frac{C(T)}{V_m^2} + \frac{D(T)}{V_m^3} + \cdots \tag{1.9}$$

Another very useful series for representing the properties of real gases is:

$$PV_m = RT + \beta P + \gamma P^2 + \delta P^3 + \cdots \tag{1.10}$$

$$\beta = B \qquad \gamma = \frac{C - B^2}{RT} \qquad \delta = \frac{D - 3BC - 2B^3}{R^2 T^2} \tag{1.11}$$

Table 1.2 Equations for Virial Coefficients

	$B(T)$	$C(T)$	$D(T)$
van der Waals	$b - \dfrac{a}{RT}$	b^2	b^3
Redlich–Kwong	$b - \dfrac{a}{RT^{3/2}}$	$b^2 + \dfrac{ab}{RT^{3/2}}$	$b^3 - \dfrac{ab^2}{RT^{3/2}}$
Berthelot	$\dfrac{9RT_c}{128P_c}\left[1 - \dfrac{6T_c^2}{T^2}\right]$	—	—
Beattie–Bridgeman	$B_0 - \dfrac{A_0}{RT} - \dfrac{c}{T^3}$	$\dfrac{A_0 a}{RT} - B_0 b - \dfrac{B_0 c}{T^3}$	$\dfrac{B_0 bc}{T^3}$

Table 1.3 Beattie–Bridgeman Constants (for volume in liter, pressure in atm, $R = 0.08206$ liter atm K^{-1} mol^{-1})

Gas	A_0	a	B_0	b	$c \times 10^{-4}$
He	0.0216	0.05984	0.01400	0	0.004
Ne	0.2125	0.2196	0.02060	0	0.101
Ar	1.2907	0.02328	0.03931	0	5.99
Kr	2.4230	0.02865	0.05261	0	14.89
H_2	0.1975	−0.00506	0.02096	−0.04359	0.050
N_2	1.3445	0.02617	0.05046	−0.00691	4.20
O_2	1.4911	0.02562	0.04624	0.004208	4.80
CO_2	5.0065	0.07132	0.10476	0.07235	66.00
CH_4	2.2769	0.01855	0.05587	−0.01587	12.83
C_2H_6	5.8800	0.05861	0.09400	0.01915	90.0
C_2H_4	6.1520	0.04964	0.12156	0.03597	22.68

From: J. O. Hirschfelder, C. F. Curtiss, and R. B. Byrd, *Molecular Theory of Gases and Liquids*, 1954: New York, John Wiley & Sons.

Table 1.4 Equation of State Parameters from Critical Constants

	van der Waals	Redlich–Kwong
a	$\dfrac{27R^2T_c^2}{64P_c}$	$0.42748\,\dfrac{R^2T_c^{5/2}}{P_c}$
b	$\dfrac{RT_c}{8P_c}$	$0.086640\,\dfrac{RT_c}{P_c}$
V_c	$3b$	$3.847b$
$z_c = \dfrac{P_cV_c}{RT_c}$	$\dfrac{3}{8}$	$\dfrac{1}{3}$

See Figure 1.12 on page 8.

$$n^* \equiv \frac{N}{V} \tag{1.15a}$$

The square root of this quantity is the **root-mean-square** (rms) speed, for which we shall use the symbol u:

$$u \equiv \langle v^2 \rangle^{1/2} \quad u = \sqrt{\frac{3RT}{M}} \tag{1.20}$$

$$\text{ave. kinetic energy} = \tfrac{3}{2}k_b T \tag{1.21}$$

The average value of x is:
$$\langle x \rangle = \int_{-\infty}^{\infty} xf(x)\, dx \tag{1.26}$$

The mean square average of x is:

$$\langle x^2 \rangle = \int_{-\infty}^{\infty} x^2 f(x)\, dx \tag{1.27}$$

The probability (P) of finding $x_1 < x < x_2$ is:

$$P(x_1, x_2) = \int_{x_1}^{x_2} f(x)\, dx \tag{1.28}$$

$$Z_{\text{wall}} = n^*\left(\frac{RT}{2\pi M}\right)^{1/2} \tag{1.31}$$

$$P = \mu\left(\frac{2\pi RT}{M}\right)^{1/2} \tag{1.32}$$

$$F(v)\, dv = 4\pi\left(\frac{m}{2\pi k_b T}\right)^{3/2} e^{-mv^2/2k_b T} v^2\, dv \tag{1.33}$$

$$v_p = \sqrt{\frac{2k_b T}{m}} \tag{1.34}$$

$$\bar{v} = \sqrt{\frac{8RT}{\pi M}} \tag{1.35}$$

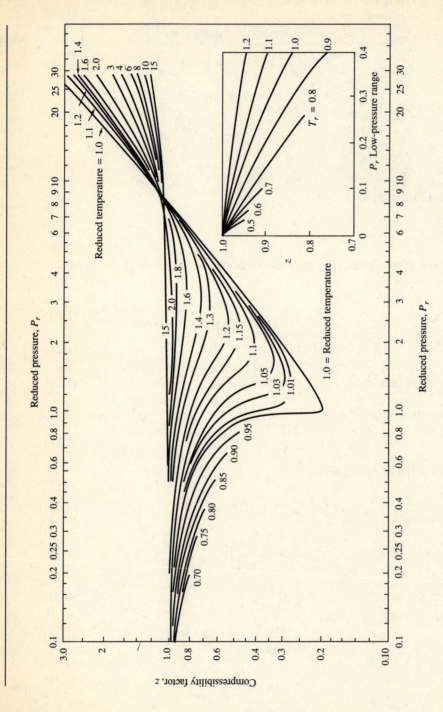

Figure 1.12 Compressibility factor. The compressibility factor as a function of reduced pressure and temperature; curves were constructed from average data of a variety of gases. (From O. A. Hougen and K. M. Watson, *Chemical Process Principles, Part II*, 1947: New York, John Wiley & Sons)

$$w \equiv v/\sqrt{2k_bT/m} \qquad P(v_1 \text{ to } v_2) = \frac{4}{\sqrt{\pi}} \int_{w_1}^{w_2} e^{-w^2} w^2 \, dw \tag{1.37b}$$

$$B(T) = 2\pi L \int_0^\infty (1 - e^{-U/k_bT}) r^2 \, dr \tag{1.38}$$

Lennard-Jones $\qquad U(r) = 4\epsilon \left[\left(\frac{\sigma}{r} \right)^{12} - \left(\frac{\sigma}{r} \right)^6 \right] \tag{1.40}$

$$B^* \equiv \frac{B}{b_0}, \qquad T^* \equiv \frac{T}{(\epsilon/k_b)} \quad \textbf{(1.42)} \qquad C^* \equiv C/b_0^2 \quad \textbf{(1.43)}$$

See Figure 1.21 on page 10.

Table 1.7   Intermolecular Potential Constants

	Lennard-Jones			Square-well			
	$(\epsilon/k)/K$	σ/nm	b_0/(cm³/mol)	$(\epsilon/k)/K$	σ/nm	b_0/(cm³/mol)	R
He	10.8	0.263	22.9	—	—	—	—
Ne	35.6	0.2749	26.20	19.5	0.2382	17.05	1.87
Ar	119.8	0.3405	49.79	69.4	0.3162	39.87	1.85
Kr	171	0.360	58.8	98.3	0.3362	47.93	1.85
H_2	29.2	0.287	29.8	—	—	—	—
N_2	95.0	0.3698	63.78	53.7	0.3299	45.29	1.87
O_2	117.5	0.358	57.9	—	—	—	—
CO	100.2	0.3763	67.21	—	—	—	—
CO_2	189	0.4486	113.9	119	0.3917	75.80	1.83
CH_4	148.2	0.3817	70.14	—	—	—	—
C_2H_6	243	0.3954	77.97	224	0.3535	55.72	1.652
C_2H_4	200	0.452	116	222	0.3347	47.29	1.667
C_6H_6	440	0.527	185	—	—	—	—

Source: J. O. Hirschfelder, C. F. Curtiss, and R. B. Bird, *Molecular Theory of Gases and Liquids,* 1954: New York, John Wiley & Sons.

square-well $\qquad B(T) = b_0[1 - (R^3 - 1)(e^{\epsilon/kT} - 1)] \tag{1.44}$

$$a = \left[\sum_i X_i \sqrt{a_i} \right]^2 \qquad b = \sum_i X_i b_i \tag{1.47}$$

$$\alpha \equiv \frac{1}{V} \left(\frac{\partial V}{\partial T} \right)_P \tag{1.49}$$

$$\kappa_T \equiv -\frac{1}{V} \left(\frac{\partial V}{\partial P} \right)_T \tag{1.50}$$

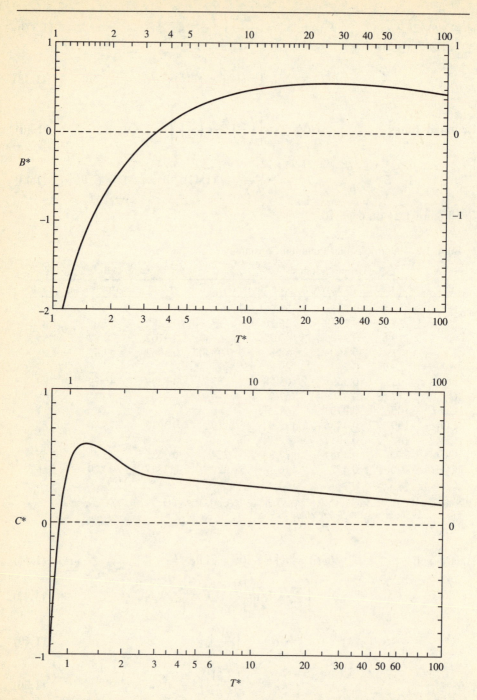

Figure 1.21 Lennard-Jones virial coefficients. (a) The second virial coefficient of a Lennard-Jones gas. (b) The third virial coefficient of a Lennard-Jones gas. These graphs are useful for estimating second and third virial coefficients for gases whose Lennard-Jones parameters have been measured.

Table 1.8 Coefficients of Thermal Expansion and Compressibility (at 20°C unless otherwise specified)

	$10^3\alpha/(\text{K}^{-1})$	$10^6\kappa_T/(\text{atm}^{-1})$	$V_m/(\text{cm}^3)$	$C_{pm}/(\text{J K}^{-1})$	$C_{vm}/(\text{J K}^{-1})$
Benzene	1.237	63.5	89.	134	92.0
CCl$_4$	1.236	91.0	97.	132	89.5
CCl$_3$H	1.273	90.0	80.2	117	—
Mercury	0.18	3.9	14.8	27.8	23.4
Ag	0.0567	0.98	10.3	25.3	
Cu (−190°C)	0.0270	0.71	—	13.3	—
(20°C)	0.0492	0.76	7.1	24.4	—
(500°C)	0.0600	0.91	—	27.5	—
Zn	0.0893	0.15	7.1	25.0	24.0
H$_2$O (0°C)	−0.0547	47	18	76.0	—
(20°C)	0.188	45	18	75.3	—
Ice (0°C)	0.11	—	19.6	37.1	—
Acetone	1.487	112.0	45.9	128.0	—
Ethyl ether	1.656	169(13°)	54.5(25°)	169(30°)	—

Data from N. A. Lange, *Handbook of Chemistry*, 9th ed., 1956: Sandusky, Ohio, Handbook Publishers, where more values can be found.

CHAPTER 2

Table 2.1 Heat Capacity of Gases (at 15°C, 1 atm, except as indicated)

	$\dfrac{C_{pm}}{(\text{JK}^{-1}\text{mol}^{-1})}$	$\dfrac{C_{vm}}{(\text{JK}^{-1}\text{mol}^{-1})}$	γ	$\dfrac{C_{pm}-C_{vm}}{R}$
He	20.88	12.55	1.664	1.002
Ar	20.93	12.59	1.668	1.003
H$_2$	28.58	20.26	1.410	1.000
N$_2$	29.04	20.69	1.404	1.004
O$_2$	29.16	20.81	1.401	1.003
Cl$_2$	34.13	25.18	1.355	1.076
CO	29.05	20.70	1.404	1.004
HCl	29.59	20.98	1.410	1.035
H$_2$S	36.11	27.36	1.320	1.053
CO$_2$	36.62	28.09	1.304	1.026
SO$_2$	40.64	31.50	1.290	1.100
N$_2$O	36.91	28.33	1.303	1.032
NH$_3$	37.29	28.47	1.310	1.060
CH$_4$ (15°)	35.46	27.07	1.31	1.01
CH$_4$ (−74°)	33.4	24.8	1.35	1.03
CH$_4$ (−115°)	30.2	21.4	1.41	1.06
C$_2$H$_2$ (acetylene)	41.73	31.16	1.260	1.037
C$_2$H$_4$ (ethylene)	42.17	33.56	1.255	1.036
C$_2$H$_6$ (ethane)	48.58	39.46	1.22	1.096

$$\Delta U \equiv U_2 - U_1 = q + w \tag{2.8}$$

$$w = -\int_{x_1}^{x_2} \frac{F}{A} A \, dx$$

$$w = -\int_{V_1}^{V_2} P_{ex} \, dV \tag{2.9}$$

$$C_v = \left(\frac{\partial U}{\partial T}\right)_V \tag{2.12}$$

$$\left(\frac{\partial U}{\partial V}\right)_T = T\left(\frac{\partial P}{\partial T}\right)_V - P \tag{2.13}$$

$$H \equiv U + PV \tag{2.17}$$

$$C_p = \left(\frac{\partial H}{\partial T}\right)_P \tag{2.20}$$

$$\left(\frac{\partial H}{\partial P}\right)_T = V - T\left(\frac{\partial V}{\partial T}\right)_P \tag{2.21}$$

ideal gas $$C_{pm} = C_{vm} + R \tag{2.24}$$

See Table 2.2 on page 13.

Table 2.3 Heat Capacities of Gases (range: bp, or 300 K to 1500 K)

Gas	$C_{pm}^{\ominus}/(\text{JK}^{-1}\text{mol}^{-1}) = a' + b'T + c'T^2 + d'T^3$			
	a'	$10^3 b'$	$10^6 c'$	$10^9 d'$
Acetone	8.468	269.45	−143.45	29.63
n-Butane	−0.050	387.04	−200.82	40.61
Ethane	5.351	177.67	−68.70	8.514
Ethanol	14.97	208.56	−71.090	—
Propane	−5.058	308.50	−161.78	33.31
Sulfur dioxide	25.72	57.923	−38.09	8.606
Sulfur trioxide	15.10	151.92	−120.62	36.19

Source: *Tables of Physical and Chemical Constants,* 13th ed., 1966: New York, John Wiley & Sons.

$$\frac{\partial C_p}{\partial P} = \frac{\partial}{\partial T}\left(V - T\frac{\partial V}{\partial T}\right) = \frac{\partial V}{\partial T} - \frac{\partial V}{\partial T} - T\frac{\partial^2 V}{\partial T^2}$$

Therefore:

$$\left(\frac{\partial C_p}{\partial P}\right)_T = -T\left(\frac{\partial^2 V}{\partial T^2}\right)_P \tag{2.25}$$

Table 2.2   Molar Heat Capacity at Constant Pressure (units: $JK^{-1} mol^{-1}$)

$C^{\ominus}_{pm} = c_1 + c_2 T + c_3 T^2 + c_4/T^2$			
Gases[a]	c_1	$10^3 c_2$	$10^6 c_3$	$10^{-5} c_4$
H_2	26.36	4.35	−0.245	1.15
O_2	29.30	6.14	−0.88	−1.59
N_2	25.79	8.09	−1.46	0.88
CO	26.28	8.055	−1.49	0.45
CO_2	41.58	15.6	−2.95	−7.97
H_2O	26.06	17.7	−2.63	2.20
Cl_2	36.76	1.02	−0.0966	−2.78
HCl	25.36	7.38	−1.15	1.54
Br_2	37.32	0.50	—	−1.26
HBr	25.32	8.18	−1.38	1.31
HI	25.89	8.43	−1.49	0.73
I_2	37.40	0.59	—	−0.71
NH_3	29.75	25.10	—	−1.55
CH_4	24.87	55.8	−10.24	−4.87
C_2H_2	47.18	25.91	−4.23	−9.37
C_2H_4	44.28	60.1	−11.1	−16.8
H_2S	32.68	12.38	—	−1.92
Liquids[b]	c_1	$10^3 c_2$	$10^6 c_3$	$10^{-5} c_4$
I_2	80.33	—	—	—
H_2O	75.48	—	—	—
$C_{10}H_8$	79.50	407.5	—	—
Hg	28.60	−6.65	6.48	0.70
Solids[c]	c_1	$10^3 c_2$	$10^6 c_3$	$10^{-5} c_4$
Al	20.67	12.38	—	—
Al_2O_3	114.8	12.8	—	−35.4
MgO	42.59	7.28	—	6.19
Cu	22.64	6.28	—	—
Pb	22.13	11.72	—	0.96
Zn	22.38	10.04	—	—
I_2	40.12	49.79	—	—
NaCl	45.94	16.32	—	—
$CaCO_3$	104.5	21.9	—	25.9
Ag	21.3	8.54	—	−1.5
AgCl	62.26	4.18	—	−11.3
graphite	14.22	9.22	−1.87	−7.51

[a] For gases, equation is valid from 298 to 3000 K, unless c_3 is missing, then to 2000 K.

[b] From melting point to boiling point.

[c] From 298 to melting point or 1500 K, except graphite, which is from 298 to 3000 K.

$$C_p = C_v + T\left(\frac{\partial V}{\partial T}\right)_P\left(\frac{\partial P}{\partial T}\right)_V \tag{2.27}$$

$$\left(\frac{\partial V}{\partial T}\right)_P = V\alpha, \qquad \left(\frac{\partial P}{\partial T}\right)_V = \frac{\alpha}{\kappa_T}$$

Therefore:

$$C_p = C_v + \frac{TV\alpha^2}{\kappa_T} \tag{2.28}$$

$$U_i = \int_\infty^V \left[T\left(\frac{\partial P}{\partial T}\right)_T - P\right] dV \tag{2.31}$$

$$H_i = \int_0^P \left[V - T\left(\frac{\partial V}{\partial T}\right)_P\right] dP \tag{2.33}$$

$$H_i = U_i + PV_m - RT \tag{2.34}$$

$$(-w)_{max} = nRT \int_{V_1}^{V_2} \frac{dV}{V}$$

$$(-w)_{max} = nRT \ln\left(\frac{V_2}{V_1}\right) \tag{2.36}$$

$$\int_{T_1}^{T_2} \frac{C_{vm}}{T} dT = -R \ln \frac{V_2}{V_1} \tag{2.37}$$

$$C_{pm} \frac{dT}{T} = R\frac{dP}{P} \tag{2.39a}$$

$$\gamma \equiv \frac{C_p}{C_v} \tag{2.40}$$

$$c = \sqrt{\frac{\gamma RT}{M}} \tag{2.41}$$

$$P_2 V_2^\gamma = P_1 V_1^\gamma \tag{2.42}$$

$$\mu = \left(\frac{\partial T}{\partial P}\right)_H \tag{2.45}$$

$$\mu = \frac{T\left(\frac{\partial V}{\partial T}\right)_P - V}{C_p} \tag{2.47}$$

Table 2.4 Critical, Boyle, and Joule–Thomson Inversion Temperatures of Some Gases

Gas	T_c/K	T_B/K	T_i/K	Gas	T_c/K	T_B/K	T_i/K
He	5.2	24.1	44.8	Ne	44.5	120	231.4
H_2	33.2	107.3	195	CO	134	342	—
N_2	126.0	324	621	O_2	154.3	423	764
Ar	150.8	410	723	CH_4	190.6	509.7	967.8
Air	132.5	347	603	CO_2	304.2	650	~1500

CHAPTER 3

$$-w = R(T_2 - T_1)\ln \frac{V_2}{V_1} \tag{3.1a}$$

$$q_2 = RT_2 \ln \frac{V_2}{V_1} \tag{3.1b}$$

$$q_1 = -RT_1 \ln \frac{V_2}{V_1} \tag{3.1c}$$

$$dS \equiv \frac{dq_{\text{rev}}}{T} \tag{3.5}$$

$$\Delta_{\text{mix}} S = -R(n_A \ln X_A + n_B \ln X_B) \tag{3.15}$$

Table 3.1 The Fundamental Equations of Thermodynamics

Definitions: $\quad H = U + PV \qquad A = U - TS \qquad G = H - TS$

Properties of matter:
$$C_v = \left(\frac{\partial U}{\partial T}\right)_V \qquad C_p = \left(\frac{\partial H}{\partial T}\right)_P \qquad \mu = \left(\frac{\partial T}{\partial P}\right)_H$$

$$\alpha = \frac{1}{V}\left(\frac{\partial V}{\partial T}\right)_P \qquad \kappa_T = -\frac{1}{V}\left(\frac{\partial V}{\partial P}\right)_T \qquad \kappa_S = -\frac{1}{V}\left(\frac{\partial V}{\partial P}\right)_S$$

Basic Equations	Maxwell Relationships	Working Equations
(a) $dU = T\,dS - P\,dV$	$\left(\frac{\partial T}{\partial V}\right)_S = -\left(\frac{\partial P}{\partial S}\right)_V$	$dU = C_v\,dT + \left[T\left(\frac{\partial P}{\partial T}\right)_V - P\right]dV$
(b) $dA = -S\,dT - P\,dV$	$\left(\frac{\partial S}{\partial V}\right)_T = \left(\frac{\partial P}{\partial T}\right)_V$	$dS = \frac{C_v}{T}\,dT + \left(\frac{\partial P}{\partial T}\right)_V dV$
(c) $dH = T\,dS + V\,dP$	$\left(\frac{\partial T}{\partial P}\right)_S = \left(\frac{\partial V}{\partial S}\right)_P$	$dH = C_p\,dT + \left[V - T\left(\frac{\partial V}{\partial T}\right)_P\right]dP$
(d) $dG = -S\,dT + V\,dP$	$\left(\frac{\partial S}{\partial P}\right)_T = -\left(\frac{\partial V}{\partial T}\right)_P$	$dS = \frac{C_p}{T}\,dT - \left(\frac{\partial V}{\partial T}\right)_P dP$

Some Derived Relationships:

$$C_v = T\left(\frac{\partial S}{\partial T}\right)_V \qquad \left(\frac{\partial G}{\partial T}\right)_P = -S \qquad \left(\frac{\partial G}{\partial P}\right)_T = V$$

$$C_p = T\left(\frac{\partial S}{\partial T}\right)_P \qquad \left(\frac{\partial A}{\partial T}\right)_V = -S \qquad \left(\frac{\partial A}{\partial V}\right)_T = -P$$

$$c = \frac{1}{\sqrt{\rho \kappa_S}} \tag{3.25}$$

$$\kappa_T = \kappa_S + \frac{TV\alpha^2}{C_p} \tag{3.27}$$

Table 3.2   Standard Entropies at 298.15 K

Solids	$S_m^{\ominus}/(\text{J K}^{-1}\,\text{mol}^{-1})$	Gases	$S_m^{\ominus}/(\text{J K}^{-1}\,\text{mol}^{-1})$	Liquids	$S_m^{\ominus}/(\text{J K}^{-1}\,\text{mol}^{-1})$
Ag	42.55	O	161.055	Br_2	152.231
AgCl	96.2	O_2	205.138	H_2O	69.91
AgBr	107.1	I	180.791	Hg	76.02
AgI	115.5	I_2	260.69	CH_3OH	126.8
Al	28.33	H	114.713	C_2H_5OH	160.7
Au	47.36	H_2	130.684	C_6H_6	173
C	5.740 (graphite)	N_2	191.61	CCl_4	216.40
C	2.377 (diamond)	Cl_2	223.066	$CHCl_3$	201.7
Cd	51.76	F_2	202.78	CH_2Cl_2	177.8
Cu	33.150	CO	197.674	C_2H_4O	153.85
Zn	41.63	CO_2	213.74	CH_3COOH	159.8
I_2	116.135	HCl	186.908		
Hg_2Cl_2	196.	NO	210.761		
K	64.18	H_2S	205.79		
Na	51.21	CH_4	186.264		
Pb	64.81	C_2H_2	200.94		
$PbSO_4$	148.57	C_2H_4	219.56		
$PbCl_2$	136.0	NH_3	192.45		
PbO_2	68.6	CCl_4	309.85		
PbO	68.70 (yellow)	$CHCl_3$	234.58		
S	31.80 (rhombic)	Ar	154.853		
S	32.6 (monoclinic)	Xe	169.683		

Source: Most data from *J. Phys. Chem. Ref. Data*, **11**, Supp. 2, 1982.

$$S(P, T) = S^{\ominus}(T) - R \ln\left[\frac{P}{P^{\ominus}}\right] + S_i(P, T) \tag{3.32}$$

$$S_i(P, T) = \int_0^P \left[\frac{R}{P} - \left(\frac{\partial V}{\partial T}\right)_P\right] dP \tag{3.33}$$

$$S(V, T) = S^{\ominus}(T) + R \ln\left[\frac{P^{\ominus} V_m}{RT}\right] + S_i(V, T) \tag{3.35}$$

$$S_i(V, T) = \int_{\infty}^V \left[\left(\frac{\partial P}{\partial T}\right)_V - \frac{R}{V}\right] dV \tag{3.36}$$

Table 3.3 Formulas for RK Gas

U_i	$\dfrac{3a}{2b\sqrt{T}}\ln\left[\dfrac{V_m}{V_m+b}\right]$
S_i	$\dfrac{a}{2bT^{3/2}}\ln\left[\dfrac{V_m}{V_m+b}\right]-R\ln\left[\dfrac{V_m}{V_m-b}\right]$
A_i	$\dfrac{a}{b\sqrt{T}}\ln\left[\dfrac{V_m}{V_m+b}\right]+RT\ln\left[\dfrac{V_m}{V_m-b}\right]$
PV_m-RT	$\dfrac{bRT}{V_m-b}-\dfrac{a}{\sqrt{T}(V_m+b)}$
H_i	U_i+PV_m-RT
G_i	A_i+PV_m-RT

Table 3.4 Properties of Rubber[a]
($T = 298$ K. Units: Pa = J m^{-3} = 10 dynes cm^{-2} = 10 ergs cm^{-3})

	Poly (isoprene) (Natural Rubber)	Butadiene-styrene (SBR, 23.5% S)	Chloroprene (Neoprene™)	Butene-isoprene (Butyl Rubber)
$\rho/(\text{g cm}^{-3})$	0.970	0.980	1.320	0.933
$\alpha/(\text{K}^{-1})$	6.6×10^{-4}	6.6×10^{-4}	$(6.1-7.2)\times10^{-4}$	5.6×10^{-4}
$c_p/(\text{J K}^{-1}\text{ g}^{-1})$	1.828	1.83	2.1–2.2	1.85
$\kappa_T/(\text{Pa}^{-1})$	5.14×10^{-10}	5.10×10^{-10}	4.40×10^{-10}	5.08×10^{-10}
ϵ max (%)[b]	750–850	400–600	800–1000	750–950
Tensile strength (Pa)[b]	$(17-25)\times10^6$	$(1.4-3.0)\times10^6$	$(25-38)\times10^6$	$(18-21)\times10^6$
Initial modulus (Pa)[c]	1.3×10^6	1.6×10^6	1.6×10^6	1.0×10^6

[a]Properties listed are for vulcanized gum stock without fillers.
[b]Break point.
[c]Initial slope of stress-strain curve.

CHAPTER 4

$$\left[\frac{\partial(G/T)}{\partial T}\right]_P = -\frac{H}{T^2} \qquad\qquad \textbf{(4.6a)}$$

$$\left[\frac{\partial(G/T)}{\partial(1/T)}\right]_P = H \qquad\qquad \textbf{(4.6b)}$$

$$\frac{dP}{dT} = \frac{\Delta_\phi H}{T\,\Delta_\phi V} \qquad\qquad \textbf{(4.19)}$$

Table 4.2   Heats of Vaporization and Fusion (values are at normal (1 atm) boiling or freezing points)

Substance	T_f/K	$\Delta_f H/(kJ/mol)$	T_b/K	$\Delta_v H/(kJ/mol)$	$\Delta_v S/(J\ K^{-1}mol^{-1})$
He	—	—	4.21	0.084	20
H_2	13.95	0.12	20.38	0.904	44.4
N_2	63.14	0.720	77.33	5.577	72.12
O_2	54.39	0.444	90.18	6.820	75.63
SO_2	197.67	7.40	263.13	24.92	94.71
CO_2			(Sublimes 194.65 K, $\Delta H \cong 25$ kJ/mole)		
NH_3	195.39	5.653	239.72	23.33	97.32
$CHClF_2$	113	—	232.4	20.23	87.05
$CClF_3$	92	—	191.8	15.50	80.81
CCl_2F_2	115	—	243.36	19.97	82.06
CCl_3F	162	—	296.92	25.00	84.20
$CHCl_3$	209.7	9.2	334.4	29.4	87.8
CCl_4	250.3	2.5	349.9	30.0	85.7
CH_4	90.67	0.941	111.66	8.18	73.3
C_2H_6	89.88	2.86	184.52	14.72	79.8
C_3H_8	85.44	3.52	231.03	18.78	81.3
$n\text{-}C_4H_{10}$	134.80	4.66	272.65	22.39	82.1
$n\text{-}C_6H_{14}$	177.80	13.03	341.89	28.85	84.4
$n\text{-}C_8H_{18}$	216.36	20.74	398.81	34.98	87.7
CH_3OH	175.4	3.17	337.9	35.27	104
C_2H_5OH	156	5.02	351.7	38.58	110
H_2O	273.15	6.01	373.15	40.66	109

$$\frac{d(\ln P)}{dT} = \frac{\Delta_v H}{RT^2}; \qquad \frac{d(\ln P)}{d(1/T)} = -\frac{\Delta_v H}{R} \qquad \textbf{(4.20)}$$

$$dw = \gamma\, d\mathscr{A} \qquad \textbf{(4.25)}$$

Table 4.3   Surface Tensions of Some Liquids (units: dyne cm^{-1} = 10^{-3}N m^{-1})

Substance	°C	γ	Substance	°C	γ
Platinum	2273	1900	Water	0	75.7
Copper	1404	1100	Water	20	72.75
Aluminum	700	840	Water	25	72.0
Lead	350	453	Water	40	69.6
Mercury	20	472	Water	60	66.2
Acetone	20	23.7	Water	80	62.6
Benzene	20	28.88	Water	100	58.8
Chloroform	20	27.14	Oxygen	−183	13.1
Ethanol	10	23.6	Oxygen	−203	18.3
Ethanol	20	22.8	Argon	−188	13.2
Ethanol	30	21.9	Argon	−183	11.9
Methanol	0	24.5	Nitrogen	−203	10.5
Methanol	20	22.6	Nitrogen	−193	8.3
Methanol	50	20.2	Nitrogen	−183	6.2

$$S_s \equiv \left(\frac{\partial S}{\partial \mathcal{A}}\right)_{T, V} = -\left(\frac{\partial \gamma}{\partial T}\right)_{V, \mathcal{A}} \tag{4.29}$$

$$\gamma = U_s + T\left(\frac{\partial \gamma}{\partial T}\right)_V$$

$$U_s = \gamma - T\left(\frac{\partial \gamma}{\partial T}\right)_V \tag{4.30}$$

$$\Delta P = \frac{4\gamma}{r} \tag{4.32}$$

$$\ln \frac{P}{P^\circ} = \frac{2\gamma M}{\rho r R T} \tag{4.36}$$

$$\mu_\alpha^\ominus + V_{m\alpha}(P - P^\ominus) = \mu_\beta^\ominus + V_{m\beta}(P - P^\ominus) \tag{4.37}$$

CHAPTER 5

$$S = k_b \ln W \tag{5.1}$$

$$P(N, p) = \frac{1}{2^N} \frac{N!}{p! \, (N - p)!} \tag{5.3}$$

$$\ln(N!) = N \ln N - N \tag{5.5}$$

$$\ln(N!) = \frac{1}{2} \ln(2\pi) + \left(N + \frac{1}{2}\right)\ln(N) - N + \frac{1}{12N} - \frac{1}{360N^3} \cdots \tag{5.6}$$

fluctuation
$$p = \frac{W}{W^{eq}} = e^{\Delta S/k_b} \tag{5.7}$$

$$W(N, \{N_i\}) = \frac{N!}{N_1! \, N_2! \, N_3! \cdots} \tag{5.8}$$

$$\Delta_{mix}S = -R \sum_i n_i \ln X_i \tag{5.10}$$

$$z = \sum_{levels} g_i e^{-\epsilon_i/k_b T} \tag{5.23}$$

$$p_i = \frac{N_i}{N} = \frac{1}{z} g_i \exp\left(-\frac{\epsilon_i}{k_b T}\right) \tag{5.24}$$

$$U - U_0 = RT^2\left(\frac{\partial \ln z}{\partial T}\right)_V \tag{5.26}$$

$$S = R \ln z + \frac{U - U_0}{T} \tag{5.28}$$

translation

$$z = \frac{(2\pi m k_b T)^{3/2} V}{h^3} \tag{5.31}$$

$$S_{\text{tr}}^{\ominus} = R \ln\left[\frac{(2\pi M)^{3/2} R^{5/2} T^{5/2}}{L^4 P^{\ominus} h^3}\right] + \frac{5}{2}R \tag{5.35}$$

$$z_{\text{vib}} = \frac{1}{1 - e^{-\theta_v/T}} \tag{5.39}$$

linear

$$z_{\text{rot}} = \frac{T}{\sigma \theta_r} \tag{5.43}$$

nonlinear

$$z_{\text{rot}} = \frac{\sqrt{\pi}}{\sigma} \frac{T^{3/2}}{\sqrt{\theta_a \theta_b \theta_c}} \tag{5.44}$$

Table 5.1 Molecular Constants for Rotation and Vibration

Linear	θ_r	θ_v (Degeneracy)
H_2	87.5	6323.8
HF	30.127	5954.5
DF	15.837	4313.9
HCl	15.238	4301.6
HBr	12.19	3812
HI	9.43	3323
N_2	2.89	3395
CH*	20.8	4117
CO	2.78	3122
NO*	2.50	2739
O_2*	2.08	2274
OH*	27.15	5374
Cl_2	0.351	813
Br_2	0.116	470
ICl	0.164	553
HCN	2.1268	3306, 1024 (2), 4765
CO_2	0.56166	1997, 960 (2), 3380
OCS	0.29188	1235.89, 753.91 (2), 2969.6
HC≡CH	1.6932	879.9 (2), 1049 (2), 2840, 4722, 4854

Nonlinear	σ	$\theta_a \theta_b \theta_c$	θ_v (degeneracy)
H_2O	2	11330	5254, 2295, 5404
CH_4	12	435.6	4193, 2196 (2), 4345 (3), 1879 (3)
CH_2Cl_2	2	0.03496	406, 1032, 1091, 1292, 1659, 1824, 2111, 4315, 4374
NH_3	3	1875.8	4801, 1367, 4912 (2), 2342 (2)
NO_2‡	2	4.24244	1954, 1089, 2396
NOCl	1	0.28667	2590, 855.9, 476.9
O_3	2	1.86061	1597.02, 1014.33, 1500.63
SO_2	2	0.61047	1656, 754, 1958
SiF_4	12	0.008235	1151, 374 (2), 1470 (3), 604 (3)

*See Table 5.2 for electronic states.
‡Ground electronic state, $g_0 = 2$.

Table 5.2 Electronic Energy Levels
Only lower levels are listed. Number in parentheses is degeneracy.

Atoms	g_0	θ_e/K (gas)
Br	4	5302.18 (2) . . .
C	1	23.60 (3), 62.44 (5), 14665 (5) . . .
Cl	4	1269.5 (2) . . .
F	4	581.403 (2) . . .
I	4	10939.1 (2) . . .
N	4	27658.7 (6), 27671.7 (4), 41492.4 (6) . . .
Na	2	24396 (2), 24420 (4) . . .
O	5	227.705 (3), 326.594 (1), 22830 (5) . . .

Molecules		
CH	2	25.8 (2), 6474? (4), 33307 (2) . . .
CH_2	3	3741 (1), 13956 (1), 43594 (1) . . .
CH_3	2	66478 (2) . . .
NO	2	174.2 (2) . . .
O_2	3	11392 (2), 18984 (1) . . .
OH	2	201.0 (2) . . .

Table 5.3 Thermodynamic Properties of Ideal Gases (per mole)

	Translation[a]	Rotation[b]	Vibration[c]
$C_v^{\ominus}$	$1.5R$	R ------ $1.5R$	$\dfrac{Ru^2e^u}{(e^u-1)^2}$
$U^{\ominus}-U_0$	$1.5RT$	RT ------ $1.5RT$	$\dfrac{R\theta_v}{e^u-1}$
$S^{\ominus}$	$1.5R\ln M + 2.5R\ln T$ $-1.15167R$	$R+R\ln z_r$ ------ $1.5R+R\ln z_r$	$\dfrac{Ru}{e^u-1} - R\ln(1-e^{-u})$
ϕ°	$1.5R\ln M + 2.5R\ln T$ $-3.65167R$	$R\ln z_r$ ------ $R\ln z_r$	$-R\ln(1-e^{-u})$

[a] M is molecular weight in g/mol.
[b] Linear above dashed line, nonlinear below.

Linear: $z_r = \dfrac{T}{\sigma\theta_r}$; nonlinear: $z_r = \dfrac{\sqrt{\pi}T^{3/2}}{\sigma(\theta_a\theta_b\theta_c)^{1/2}}$.

[c] $u = \theta_e/T$. Sum over all vibrations.

$$\phi^{\circ} \equiv -\frac{(G^{\ominus}-H_0)}{T} \tag{5.51}$$

$$\phi' = -\frac{G^{\ominus}-H_{298}^{\ominus}}{T} \tag{5.55}$$

$$\phi' = \phi^{\circ} + \frac{H_{298}^{\ominus}-H_0}{T} \tag{5.57}$$

$$\phi'(T) = S^{\ominus}(298.15) + \int_{298.15}^{T}\frac{C_p}{T}\,dT - \frac{1}{T}\int_{298.15}^{T}C_p\,dT \tag{5.58}$$

CHAPTER 6

Table 6.1   Thermodynamic Properties at 298.15 K, $P^{\ominus} = 0.1$ MPa[1]

	$\Delta_f H^{\ominus}/$(kJ/mol)	$\Delta_f G^{\ominus}/$(kJ/mol)	$S^{\ominus}/$(JK^{-1}/mol^{-1})	$C_p/$(JK^{-1}mol^{-1})
Ag(s)	0	0	42.55	25.351
Ag$_2$SO$_4$(s)	−715.88	−618.41	200.4	131.38
AgBr(s)	−100.37	−96.90	107.1	52.38
Ag$_2$C$_2$O$_4$(s)	−673.2	−584.0	209.	
AgCl(s)	−127.068	−109.789	96.2	50.79
AgCN(s)	146.0	156.9	107.19	66.73
AgI(s)	−61.84	−66.19	115.5	56.82
AgSCN(s)	87.9	101.39	131.0	63.
Br(g)	111.884	82.396	175.022	20.786
Br$_2$(g)	30.907	3.110	245.463	36.02
Br$_2$(liq)	0.	0.	152.231	75.689
BrCl(g)	14.64	−0.98	240.10	34.98
C(diamond)	1.895	2.900	2.377	6.113
C(graphite)	0.	0.	5.740	8.527
CO(g)	−110.525	−137.168	197.674	29.142
CO$_2$(g)	−393.509	−394.359	213.74	37.11
COCl$_2$(g)	−218.8	−204.6	283.53	57.66
CS$_2$(g)	117.36	67.12	237.84	45.40
CS$_2$(liq)	89.70	65.27	151.34	75.7
CCl$_4$(g)	−102.9	−60.59	309.85	83.30
CCl$_4$(liq)	−135.44	−65.21	216.40	131.75
CF$_4$(g)	−925.	−879.	261.61	61.09
CH$_2$Cl$_2$(g)	−92.47	−65.87	270.23	50.96
CH$_3$Cl(g)	−80.83	−57.37	234.58	40.75
CH$_3$OH(g)	−200.66	−161.96	239.81	43.89
CH$_3$OH(liq)	−238.66	−166.27	126.8	81.6
CH$_4$(g)	−74.81	−50.72	186.264	35.309
CHCl$_3$(liq)	−134.47	−73.66	201.7	113.8
HCOOH(liq)	−424.72	−361.35	128.95	99.04
C$_2$H$_2$(g)	226.73	209.20	200.94	43.93
C$_2$H$_4$(g)	52.26	68.15	219.56	43.56
C$_2$H$_4$Cl$_2$(g)[2]	−129.79	−73.87	308.39	78.7
C$_2$H$_4$O(g)[3]	−52.63	−13.01	242.53	47.91
C$_2$H$_5$OH(g)	−235.10	−168.49	282.70	65.44
C$_2$H$_5$OH(liq)	−277.69	−174.78	160.7	111.46
C$_2$H$_6$(g)	−84.68	−32.82	229.60	52.63
CaCl$_2$(s)	−795.8	−748.1	104.6	72.59
CaCO$_3$(s, ara)[4]	−1207.13	−1127.75	88.7	81.25
CaCO$_3$(s, cal)[5]	−1206.92	−1128.79	92.9	81.88
CaF$_2$(g)	−781.6	−790.4	274.37	51.25
CaF$_2$(s)	−1219.6	−1167.3	68.87	67.03
CaO(s)	−635.09	−604.03	39.75	42.80
Cl(g)	121.679	105.680	165.198	21.840
Cl$_2$(g)	0.	0.	223.066	33.907

Table 6.1 (Continued)

	$\Delta_f H^{\ominus}/(\text{kJ/mol})$	$\Delta_f G^{\ominus}/(\text{kJ/mol})$	$S^{\ominus}/(\text{JK}^{-1}\text{mol}^{-1})$	$C_p/(\text{JK}^{-1}\text{mol}^{-1})$
$ClF(g)$	−54.48	−55.94	217.89	32.05
$ClF_3(g)$	−163.2	−123.0	281.61	63.85
$ClF_3 \cdot HF(g)$	−450.6	−384.0	360.	
$ClO(g)$	101.84	98.11	226.63	31.46
$ClO_2(g)$	102.5	120.5	256.84	41.97
$ClO_3F(g)$	−23.8	48.2	278.97	64.94
$F(g)$	78.99	61.91	158.754	22.744
$F_2(g)$	0.	0.	202.78	31.30
$H(g)$	217.965	203.247	114.713	20.784
$H_2(g)$	0	0	130.684	28.824
$H_2O(g)$	−241.818	−228.572	188.825	33.577
$H_2O(liq)$	−285.830	−237.129	69.91	75.291
$H_2S(g)$	−20.63	−33.56	205.79	34.23
$HF(g)$	−271.1	−273.2	173.779	29.133
$HCl(g)$	−92.307	−95.299	186.908	29.12
$HBr(g)$	−36.40	−53.45	198.695	29.142
$HI(g)$	26.48	1.70	206.594	29.158
$Hg_2Br_2(s)$	−206.90	−181.075	218.	
$Hg_2Cl_2(s)$	−265.22	−210.745	192.5	
$Hg_2I_2(s)$	−121.34	−111.	233.5	
$Hg_2SO_4(s)$	−743.12	−625.815	200.66	131.96
$HgCl_2(s)$	−224.3	−178.6	146.0	
$HgO(s, red)$	−90.83	−58.539	70.29	44.06
$I(g)$	106.838	70.250	180.791	20.786
$I_2(g)$	62.438	19.327	260.69	36.90
$I_2(s)$	0.	0.	116.135	54.438
$IF_5(g)$	−822.49	−751.73	327.7	99.2
$K(g)$	89.24	60.59	160.336	20.786
$K(s)$	0.	0.	64.18	29.58
$KCl(g)$	−214.14	−233.0	239.10	36.48
$KCl(s)$	−436.747	−409.14	82.59	51.30
$N(g)$	472.704	455.563	153.298	20.786
$N_2(g)$	0	0	191.61	29.06
$NH_3(g)$	−46.11	−16.45	192.45	35.06
$NO(g)$	90.25	86.55	210.761	29.844
$NO_2(g)$	33.18	51.31	240.06	37.20
$N_2H_4(g)$	95.40	159.35	238.47	49.58
$N_2H_4(liq)$	50.63	149.34	121.21	98.87
$N_2O(g)$	82.05	104.20	219.85	38.45
$N_2O_3(g)$	83.72	139.46	312.28	65.61
$N_2O_4(g)$	9.16	97.89	304.29	77.28
$N_2O_4(liq)$	11.3	115.1	355.7	142.7
$N_2O_5(g)$	−19.50	97.54	209.2	84.5
$Na(g)$	107.32	76.761	153.712	20.786

(*continued*)

Table 6.1   (Continued)

	$\Delta_f H^{\ominus}/(\text{kJ/mol})$	$\Delta_f G^{\ominus}/(\text{kJ/mol})$	$S^{\ominus}/(\text{JK}^{-1}\text{mol}^{-1})$	$C_p/(\text{JK}^{-1}\text{mol}^{-1})$
Na(s)	0.	0.	51.21	28.24
NaCl(g)	−176.65	−196.66	229.81	35.77
NaCl(s)	−411.153	−384.138	72.13	50.50
O(g)	249.170	231.731	161.055	21.912
O_2(g)	0.	0.	205.138	29.355
O_3(g)	142.7	163.2	238.93	39.20
$PbBr_2$(s)	−278.7	−261.92	161.5	80.12
$PbCl_2$(s)	−359.41	−314.10	136.0	
PbI_2(s)	−175.48	−173.64	174.85	77.36
PbO(s, red)	−218.99	−188.93	66.5	45.81
PbO(s, yellow)	−217.32	−187.89	68.70	45.77
PbO_2(s)	−277.4	−217.33	68.6	64.64
$PbSO_4$(s)	−919.94	−813.14	148.57	103.207
S(s, rh)	0	0	31.80	22.64
S_2(g)	128.37	79.3	228.18	32.47
SO_2(g)	−296.830	−300.194	248.22	39.87
SO_2Cl_2(g)	−364.0	−320.0	311.94	77.0
SO_3(g)	−395.72	−371.06	256.76	50.67
$SOCl_2$(g)	−212.5	−198.3	309.77	66.5
SF_4(g)	−774.9	−731.3	292.03	73.01
SF_6(g)	−1209.	−1105.3	291.82	97.28
Si(s)	0.	0.	18.83	20.00
$SiCl_4$(g)	−657.01	−616.98	330.73	90.25
$SiCl_4$(liq)	−687.	−619.84	239.7	145.31
SiF_4(g)	−1614.94	−1572.65	282.49	73.64
SiH_4(g)	34.3	56.9	204.62	42.84
SiN(g)	486.52	456.08	216.76	30.17
SiS(g)	112.47	60.89	223.66	32.26
SiO_2(s, am)[6]	−903.49	−850.70	46.9	44.4
SiO_2(s, cry)[7]	−909.48	−855.43	42.68	44.18
SiO_2(s, qtz)[8]	−910.94	−856.64	41.84	44.43
SiO_2(s, tri)[9]	−909.06	−855.26	43.5	44.60
TlCl(s)	−204.14	−184.92	111.25	50.92

1. *Source:* The NBS Tables of Chemical Thermodynamic Properties, *J. Phys. Chem. Ref. Data,* 11, Supplement No. 2, 1982.
2. 1-2 dichloroethane
3. Ethylene oxide
4. Aragonite
5. Calcite
6. amorphous
7. Cristobalite
8. Quartz
9. Tridymite

Table 6.2 Heats of Combustion at 298.15 K*

Substance	Formula	$\Delta_c H^\ominus$/(kJ/mol)
Graphite	C	−393.51
Diamond	C	−395.39
Methane	CH_4	−890.4
Ethane	C_2H_6	−1559.8
Propane	C_3H_8	−2220.0
n-Butane	C_4H_{10}	−2878.5
n-Octane	C_8H_{18}	−5452
Methanol	CH_3OH	−726.1
Ethanol	C_2H_5OH	−1367
Benzoic acid	C_6H_5COOH	−3226.7
Sucrose	$C_{12}H_{22}O_{11}$	−5643.8
Thiophene	C_4H_4S	−2805

* For one mole of substance with $CO_2(g)$, $SO_2(g)$, and $H_2O(liq)$ as products.

$$\Delta_{\text{rxn}} H(T_2) = \Delta_{\text{rxn}} H(T_1) + \int_{T_1}^{T_2} \Delta_{\text{rxn}} C_p \, dT \qquad (6.8)$$

$$\Delta_{\text{rnx}} H^{\ominus} + \int_{T_1}^{T_2} C_p^{\text{prod}} \, dT = 0 \qquad (6.9)$$

$$\Delta_{\text{rxn}} G^{\ominus} = -RT \ln K_a \qquad (6.15)$$

$$\ln K_a = \frac{\Delta \phi^{\circ}}{R} - \frac{\Delta H_0^{\ominus}}{RT} \qquad (6.18)$$

Table 6.4   Free-Energy Functions (based on 0 K, $P^{\ominus} = 1$ atm)

	$\phi^{\circ} = -\dfrac{G_m^{\ominus} - H_0}{T}$ (JK^{-1}mol^{-1})					$H_m^{\ominus}(298.15) - H_0$ (kJ/mol)	$\Delta_f H_0^{\ominus}$ (kJ/mol)
	298.15	500	1000	1500	2000		
Graphite	2.2	4.85	11.6	17.5	22.5	1.050	0
H(g)	93.81	104.6	114.8	127.4	133.4	6.197	216.0
H$_2$(g)	102.2	116.9	137.0	148.9	157.6	8.468	0
H$_2$O(g)	155.5	172.8	196.7	211.7	223.1	9.908	−238.94
O(g)	138.4	150.0	165.1	173.8	179.8	6.724	246.2
O$_2$(g)	176.0	191.0	212.1	225.1	234.7	8.661	0
O$_3$(g)	204.1	222.9	251.8	270.7	284.5	10.36	145
N(g)	132.4	143.2	157.6	166.0	172.0	6.197	470.87
N$_2$(g)	162.4	177.5	197.9	210.4	219.6	8.669	0
NO(g)	179.8	195.6	217.0	230.0	239.5	9.180	89.87
N$_2$O(g)	187.8	205.5	233.3	252.2	—	9.586	84.98
NO$_2$(g)	205.8	224.3	252.0	270.2	284.0	10.31	36.32
CO(g)	168.4	183.5	204.1	216.6	225.9	8.673	−113.81
CO$_2$(g)	182.3	199.5	226.4	224.7	258.8	9.364	−393.17
CH$_4$(g)	152.5	170.5	199.4	221.1	239	10.03	−66.90
CH$_2$O(g)	185.1	203.1	230.6	250.6	266.0	10.01	−112
CH$_3$OH(g)	201.4	222.3	257.7	—	—	11.43	−190.2
C$_2$H$_5$OH(g)	235.1	262.8	315.0	356.3	—	14.2	−219.3
CH$_3$CHO(g)	221.1	245.5	288.8	—	—	12.84	−155.4
CH$_3$COOH(g)	236.4	264.6	317.6	357.1	—	13.8	−420.49
C$_2$H$_2$(g)	167.3	186.2	230.2	239.5	256.6	10.01	227.3
C$_2$H$_4$(g)	184.0	203.9	239.7	267.5	290.6	10.56	60.75
C$_2$H$_6$(g)	189.4	212.4	255.7	290.6	—	11.95	−69.12
NH$_3$(g)	159.0	176.9	203.5	221.9	236.6	9.92	−39.2
Cl$_2$(g)	192.2	208.6	231.9	246.2	256.6	9.2	0
COCl$_2$(g)	240.6	266.2	304.6	331.1	351.1	12.87	−217.8
CH$_3$Cl(g)	198.5	217.8	251.1	274.2	—	10.41	−74.1
CH$_2$Cl$_2$(g)	230.5	252.5	291.1	318.17	—	11.86	−79
CHCl$_3$(g)	248.1	275.3	321.2	353.0	—	14.18	−96
C$_3$H$_6$(g)	221.5	248.2	299.4	340.7	—	13.54	35.4
C$_3$H$_8$(g)	220.6	250.2	310.0	359.2	—	14.69	−81.50

Source: G. N. Lewis, M. Randall, K. S. Pitzer, and L. Brewer, *Thermodynamics,* 2nd ed. (New York; McGraw-Hill Book Co., 1961).

Table 6.5   Free-Energy Functions (based on 298.15 K, $P^\ominus = 1$ atm)

	$\phi' = -\dfrac{G_m^\ominus - H_{m,298.15}^\ominus}{T}$ (J/K)				$\Delta_f H_{298.15}^\ominus$
	298.15 K	500 K	1000 K	1500 K	(kJ)
$Fe_3C(s)$	101	114.1	160.1	196.2	+21
$Cr_{23}C_6(s)$	610.0	689.5	979.1	1128	-411
$Fe(s)$	27.2	30.2	42.1	53.8	0
$Cr(s)$	23.8	26.5	37.0	46.1	0
$Cu(s)$	33.3	36.1	46.4 melts	55.8	0
$CuO(s)$	42.6	47.8	67.3	(84.0)	-157
$Cu_2O(s)$	92.9	100.5	129.3	153.6	-170
$Fe_3O_4(s)$	146	165.5	246.7	306.3	-1120
$Fe_2O_3(s)$	87.4	100.4	152.4	196.3	-823.4
$Ti(s)$	30.5	33.5	44.3	54.2	0
$TiC(s)$	24.2	28.7	46.1	60.8	-185
$TiN(s)$	30.3	35.0	53.2	68.3	-338
$N_2(g)$	191.5	194.8	206.6	216.2	0
$O_2(g)$	205.1	208.5	220.8	230.9	0
$O_3(g)$	238.8	243.6	262.1	277.8	143
$S(s, rh.)$	31.93	34.97	46.85	58.07	0
$S_2(g)$	228.1	231.9	245.7	256.7	129.0
$S_8(g)$	430.2	448.6	516.1	569.8	101.3
$H_2S(g)$	205.6	209.6	224.5	237.3	-20.4
$SO_2(g)$	248.1	252.9	271.2	286.7	-296.8
$SO_3(g)$	256.7	262.9	267.7	308.9	-396.8

Source: G. N. Lewis, M. Randall, K. S. Pitzer, and L. Brewer, *Thermodynamics,* 2nd ed. (New York: McGraw-Hill Book Co., 1961).

$$K_a = e^{-\Delta H_0^\ominus/RT} \prod \left(\frac{z^\ominus(i)}{L}\right)^{\nu_i} \tag{6.20}$$

$$\ln K_a = \frac{\Delta \phi'}{R} - \frac{\Delta H_{298.15}^\ominus}{RT} \tag{6.21}$$

$$K_a = e^{\Delta S^\ominus/R} e^{-\Delta H^\ominus/RT} \tag{6.22}$$

$$\frac{d\ln K_a}{d(1/T)} = -\frac{\Delta_{rxn} H^\ominus}{R} \quad \text{or} \quad \frac{d\ln K_a}{dT} = \frac{\Delta_{rxn} H^\ominus}{RT^2} \tag{6.23}$$

$$f = \phi P \tag{6.28}$$

$$z = \frac{PV_m}{RT}$$

$$\ln \phi = \int_0^P \frac{z-1}{P} dP \tag{6.30}$$

See page 28 for Figure 6.8.

$$\Gamma = \exp\left(\frac{(P - P^\ominus)V_m}{RT}\right) \tag{6.31}$$

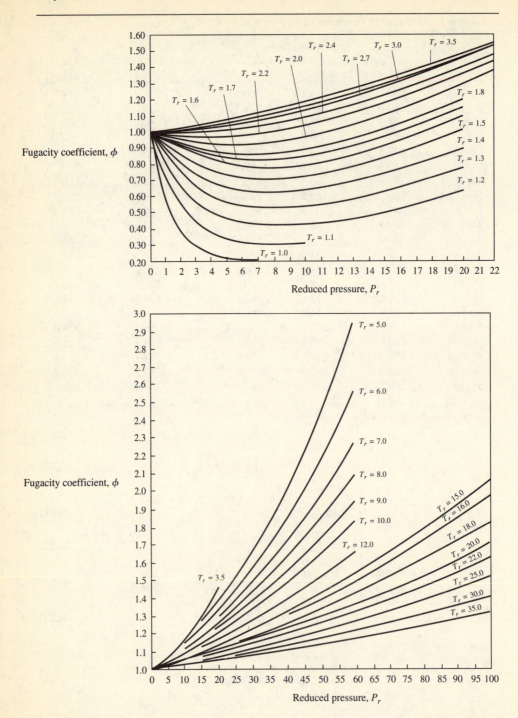

Figure 6.8 Corresponding state graphs for fugacity coefficients. (From R. H. Newton, *Ind. Eng. Chem.*, 27, 302 (1935).)

CHAPTER 7

$$X_2 = \frac{m}{m + \dfrac{1000\,g}{M_1}} \tag{7.2}$$

$$m = \frac{1000\,g}{M_1}\ \frac{X_2}{1 - X_2} \tag{7.3}$$

$$X_2 = \frac{cM_1}{1000\rho + c(M_1 - M_2)} \tag{7.4}$$

$$\bar{V_i} \equiv \left(\frac{\partial V}{\partial n_i}\right)_{T,P,n_{j\neq i}} \tag{7.5}$$

$$V = n_1 \bar{V_1} + n_2 \bar{V_2} \tag{7.9}$$

$$F = p(c - 1) + 2 - c(p - 1)$$
$$F = c + 2 - p \tag{7.12}$$

$$\mu_i - \mu_i^* = RT \ln \frac{f_1}{f_1^*} \cong RT \ln \frac{P_i}{P_i^*}$$

$$\Delta_{\text{mix}} G = RT\left(n_1 \ln \frac{P_1}{P_1^*} + n_2 \ln \frac{P_2}{P_2^*}\right) \tag{7.20}$$

Raoult's law (RL) $\qquad P_i = X_i P_i^* \tag{7.21}$

Dalton's law $\qquad P_i = Y_i P \tag{7.22}$

$$a_i = \frac{f_i}{f_i^{\ominus}} = \frac{P_i}{P_i^*} \tag{7.25}$$

$$\Delta_{\text{mix}} G^{\text{ex}} = RT(n_1 \ln \gamma_1 + n_2 \ln \gamma_2) \tag{7.27}$$

Table 7.2 Regular-Solution Constants

Mixture	T	w/J	$\dfrac{\partial w}{\partial T}$	$\dfrac{\partial^2 w}{\partial T^2}$
CCl$_4$/benzene	298	324	−0.368	−0.021
Benzene/cyclohexane	293	1275	−7.03	−0.046
CCl$_4$/cyclohexane	313	267	−0.937	
Benzene/toluene	353	−41	−0.62	0.005
CS$_2$/acetone	308	4175	−5.36	−0.043

Source: G. N. Lewis, M. Randall, K. Pitzer, and G. Brewer, *Thermodynamics,* 1961: New York, McGraw-Hill Book Co.

$$\gamma_1 = \exp\left[\frac{X_2^2 w}{RT}\right]; \qquad \gamma_2 = \exp\left[\frac{X_1^2 w}{RT}\right] \tag{7.28}$$

$$T_c = \frac{w}{2R} \tag{7.29}$$

Table 7.3 Solubility of Gases at 1 atm Pressure (25°C unless otherwise specified)

Solvent	Gas	X_2	Solvent	Gas	X_2
Benzene	Ar	8.77×10^{-4}	Water	CO_2	7.1×10^{-4} (20°C)
Cyclohexane	Ar	1.49×10^{-3}	Water	CO_2	6.1×10^{-4} (25°C)
n-Hexane	Ar	2.53×10^{-3}	CCl_4	CO_2	1.0×10^{-2} (20°C)
Water	Ar	2.7×10^{-5}	Water	CH_4	2.82×10^{-5} (19.8°C)
Water	N_2	1.2×10^{-5}	Water	C_2H_6	4.00×10^{-5} (19.8°C)
Water	O_2	2.3×10^{-5}	Water	C_3H_8	3.17×10^{-5} (19.8°C)

$$k_x = \lim_{X_2 \to 0} \left(\frac{P_2}{X_2} \right) \tag{7.31}$$

Table 7.4 Conventions for Solution Standard States

	Activity		
	Defined	Measured	Ideal State
Solvent (RL)*	$a_1 = \gamma_1 X_1$	$a_1 = \dfrac{P_1}{P_1^*}$	$\gamma_1 \to 1$ as $X_1 \to 1$
Solute (HL) (X)	$a_2 = \gamma_{2x} X_2$	$a_2 = \dfrac{P_2}{k_x}$	
(m)	$a_2 = \gamma_{2m} m$	$a_2 = \dfrac{P_2}{k_m}$	$\gamma_2 \to 1$ as $X_2 \to 0$
(c)	$a_2 = \gamma_{2c} c$	$a_2 = \dfrac{P_2}{k_c}$	

* This convention is sometimes used for solute too.

$$\mu_i = \mu_i^{\ominus} + RT \ln a_i$$

$$d(\ln a_2) = -\frac{n_1}{n_2} d(\ln a_1) \tag{7.35}$$

$$\phi = \frac{-1000(g/kg) \ln a_1}{\nu m M_1} \tag{7.36}$$

$$\ln (\gamma_{2m}) = (\phi - 1) + \int_0^m \left(\frac{\phi - 1}{m'} \right) dm' \tag{7.37}$$

Table 7.5 Activity Coefficients of Amino Acids in Water (25°C, solute standard state, molality scale)

	m					
	0.2	0.5	1.0	2.0	Sat. γ	Sat. m
Alanine	1.00	1.01	1.02	—	1.045	1.862
Alanylalanine	0.98	0.99	1.04	—	—	—
Glycine	0.960	0.908	0.875	0.787	0.729	3.37
Proline	1.02	1.05	1.10	1.21	3.13	14.1
Serine	0.95	0.89	0.81	0.70	0.602	4.02
Alanylglycine	0.93	0.87	0.86	—	—	—

$$K_f = \left[\frac{M_1 R (T_f^*)^2}{1000(\text{g/kg}) \, \Delta_{\text{fus}} H} \right]$$ (7.38)

Table 7.6   Freezing-Point Constants

Solvent	T_f^*	K_f	b
Water	273.15	1.860	4.8×10^{-4}
N-methylacetamide	303.66	5.77	1.4×10^{-3}
Benzene	278.6	5.12	—
Camphor	451.5	40.0	—
Cyclohexane	279.6	20.0	—
Ethylene carbonate	309.52	5.32	2.5×10^{-3}
Phenol	315	7.27	—

$$\theta \approx K_f m$$ (7.40)

$$\phi = \frac{\theta}{K_f m}(1 + b\theta) \quad \text{or} \quad b = \frac{1}{T_f^*} - \frac{\Delta_{\text{fus}} C_p}{\Delta_{\text{fus}} H}$$ (7.41)

$$\Pi \cong cRT$$ (7.43)

Table 7.7   Free Energies of Formation in Aqueous Solution (25°C; solute standard state, molality scale.)

	$\Delta_f G^{\ominus}(\text{pure})/\text{kJ/mol}$		$\Delta_f G^{\ominus}(\text{aq})/(\text{kJ/mol})$
Acetaldehyde	−133.7	(gas)	−139.7
Acetic acid	−392.3	(liq)	−399.5
Carbon dioxide	−394.4	(gas)	−386.2
Hydrogen sulfide	−33.0	(gas)	−27.4
Ammonia	−16.7	(gas)	−26.7
Ethanol	−174.8	(liq)	−181.5
Methanol	−166.23	(liq)	−175.23
n-Propanol	−172.4	(liq)	−175.8
iso-Propanol	−181.0	(liq)	−185.9
n-Butanol	−169.0	(liq)	−171.8
Formaldehyde	−110.0	(gas)	−130.5
Urea	−197.15	(s)	−203.84
Sucrose	−1544.7	(s)	−1551.8
α-D-Glucose	−910.27	(s)	−917.22
Fructose	—		−915.38
Glycine	−370.7	(s)	−373.0
L-Alanine	−369.9	(s)	−371.3
Alanylglycine (DL)	—		−479.36
Water	−237.191	(liq)	—

Source: Data from J.T. Edsall and J. Wyman. *Biophysical Chemistry,* Vol. I, 1958: New York, Academic Press, which uses $P^{\ominus} = 1$ atm. More data are in Tables 6.1 and 8.3.

CHAPTER 8

$$\gamma_\pm = (\gamma_+^{\nu_+} \gamma_-^{\nu_-})^{1/\nu} \tag{8.5}$$

$$a_{salt} = a_\pm^\nu = (m_+^{\nu_+} m_-^{\nu_-})\gamma_\pm^\nu \tag{8.6}$$

Table 8.1 Mean Ionic Activity Coefficients in Water (25°C)

m	KCl	NaCl	HCl	NaClO$_4$	H$_2$SO$_4$	CuSO$_4$
0.001	0.9648	0.966	0.966	—	0.830	0.74
0.005	0.927	0.929	0.928	—	0.639	0.53
0.01	0.901	0.904	0.904	—	0.544	0.41
0.02	0.868	0.875	0.875	—	0.453	0.31
0.05	0.816	0.823	0.830	—	0.340	0.21
0.1	0.769	0.778	0.796	0.775	0.265	0.16
0.2	0.718	0.735	0.767	0.729	0.209	0.11
0.5	0.649	0.681	0.757	0.668	0.154	0.068
1.0	0.603	0.673	0.809	0.629	0.130	0.047
4.0	—	0.791	1.74	—	0.172	—

m	CaCl$_2$	ZnCl$_2$	CdSO$_4$	Cr(NO$_3$)$_3$	ZnSO$_4$	LaCl$_3$
0.005	0.789	0.767	0.476	—	0.477	—
0.01	0.732	0.708	0.383	—	0.387	—
0.02	0.669	0.642	—	—	0.298	—
0.05	0.584	0.556	0.199	—	0.202	—
0.1	0.524	0.502	0.137	0.319	0.148	0.331
0.2	0.491	0.448	—	0.285	0.104	0.298
0.5	0.510	0.376	0.061	0.291	0.063	0.266
1.0	0.725	0.325	0.042	0.401	0.044	0.481

$$I \equiv \frac{1}{2} \sum_i z_i^2 m_i \tag{8.9}$$

$$\ln \gamma_\pm = \frac{-\alpha_{DA} |z_+ z_-| \sqrt{I}}{1 + Ba_0 \sqrt{I}} \tag{8.10}$$

Table 8.2 Constants for the Debye–Hückel Formula for Water

t/°C	ρ*(H$_2$O)	α$_{DH}$ (molal)	α$_{DH}$ (molal)	B (For a$_0$ in angstroms = 10^{-8} cm)
0	0.999841	1.133	1.133	0.324
10	0.999700	1.149	1.149	—
20	0.998263	1.167	1.168	—
25	0.997044	1.177	1.179	0.329
30	0.995646	1.190	1.195	—
40	0.99221	1.207	1.212	0.332
100	0.95835	1.372	1.401	—

$$\ln \gamma_\pm = \frac{-\alpha_{DH} |z_+ z_-| \sqrt{I}}{1 + \sqrt{I}} + 2\left(\frac{\nu_+^2 + \nu_-^2}{\nu_+ + \nu_-}\right)\beta m \tag{8.13}$$

Table 8.3 Thermodynamic Properties in Water at 298.15 K (ao standard state: ideal dilute solution(HL), 1 mol/kg)[a]

	$\Delta_f H^{\ominus}$/kJ/mol	$\Delta_f G^{\ominus}$/kJ/mol	$S_m^{\ominus}$/JK^{-1} mol^{-1}	$C_{pm}^{\ominus}$/JK^{-1} mol^{-1}
Cations				
H^{+b}	0.	0.	0.	0.
Ag$^+$	105.579	77.107	72.68	21.8
Be^{2+}	−382.8	379.73	−129.7	
Ca^{2+}	−542.83	−553.58	−53.1	
Cd^{2+}	−75.90	−77.612	−73.2	
Cu$^+$	71.67	49.98	40.6	
Cu^{2+}	64.77	65.49	−99.6	
Fe^{2+}	−89.1	−78.90	−137.7	
Fe^{3+}	−48.5	−4.7	−315.9	
K$^+$	−252.38	−283.27	102.5	21.8
Li$^+$	−278.49	−293.31	13.4	68.6
Mg^{2+}	−466.85	−454.8	−138.1	
Na$^+$	−240.12	−261.905	59.0	46.4
NH$_4^+$	−132.51	−79.31	113.4	79.9
CH$_3$NH$_3^+$	−124.93	−39.86	142.7	
Pb^{2+}	−1.7	−24.43	10.5	
Tl$^+$	5.36	−32.40	125.5	
Zn^{2+}	−153.89	−147.6	−112.1	46.
Anions:				
OH$^-$	−229.994	−157.244	−10.75	−148.5
F$^-$	−332.63	−278.79	−13.8	−106.7
Cl$^-$	−167.159	−131.228	56.5	−136.4
ClO$^-$	−107.1	−36.8	42.	
ClO$_2^-$	−66.5	17.2	101.3	
ClO$_3^-$	−103.97	−7.95	162.3	
ClO$_4^-$	−129.23	−8.52	182.0	
Br$^-$	−121.55	−103.96	82.4	−141.8
I$^-$	−55.19	−51.57	111.3	−142.3
I$_3^-$	−51.5	−51.4	239.3	
HS$^-$	−17.6	12.08	62.8	
S^{2-}	33.1	85.8	−14.6	
HSO$_3^-$	−626.22	−527.73	139.7	
SO$_3^{2-}$	−635.5	−486.5	−29.	
HSO$_4^-$	−887.34	−755.91	131.8	−84.
SO$_4^{2-}$	−909.27	−744.53	20.1	−293.
NO$_3^-$	−205.0	−108.74	146.4	−86.6
CH$_3$O$^-$	−193.47	−70.90	−41.4	
CN$^-$	150.6	172.4	94.1	
SCN$^-$	76.44	92.71	144.3	−40.2
OCN$^-$	−146.0	−97.4	106.7	
HCOO$^-$ (formate)	−425.55	−351.0	92.	−87.9
C$_2$O$_4^{2-}$ (oxalate)	−825.1	−673.9	45.6	
HC$_2$O$_4^-$	−818.4	−698.34	149.4	
CH$_3$COO$^-$ (acetate)	−486.01	−369.31	86.6	−6.3
HCO$_3^-$	−691.99	−586.77	91.2	
CO$_3^{2-}$	−677.14	−527.81	−56.9	
AgCl$_2^-$	−245.2	−215.4	231.4	

(continued)

Table 8.3   (Continued)

	$\Delta_f H^{\ominus}$/kJ/mol	$\Delta_f G^{\ominus}$/kJ/mol	$S_m^{\ominus}$/JK^{-1} mol^{-1}	$C_{pm}^{\ominus}$/JK^{-1} mol^{-1}
Neutrals (ao):	−000.000	−000.000	000.0	
HCOOH	−425.43	−372.3	163.	
CH$_3$OH	−245.931	−175.31	133.1	
CH$_3$COOH	−485.76	−396.46	178.7	
HCN	107.1	119.7	124.7	
CO$_2$	−413.80	−385.98	117.6	
NH$_3$	−80.29	−26.50	111.3	
CH$_3$NH$_2$	−70.17	20.77	123.4	
Cl$_2$	−23.4	6.94	121.	
HClO	−120.9	−79.9	142.	
H$_2$S	−39.1	−27.83	121.	
SO$_2$	−322.980	−300.676	161.9	
H$_2$SO$_3$	−608.81	−537.81	232.2	
I$_2$	22.6	16.40	137.2	
AgCl	−72.8	−72.8	154.0	

[a]*Source*: The NBS Tables of Chemical Thermodynamic Properties, *J. Phys. Chem. Ref. Data*, **11**, Supplement No. 2, 1982.
[b] All values are measured relative to H$^+$, which is zero by convention.

Table 8.4   Cation Transference Numbers at 25°C

Substance	Concentration/(eq/liter)					
	0.01	0.02	0.05	0.1	0.2	1.0
HCl	0.8251	0.8266	0.8292	0.8314	0.8337	
LiCl	0.3289	0.3261	0.3211	0.3168	0.3112	0.287
NH$_4$Cl	0.4907	0.4906	0.4905	0.4907	0.4911	
NaCl	0.3918	0.3902	0.3876	0.3854	0.3821	
KCl	0.4902	0.4901	0.4899	0.4898	0.4894	0.4882
KNO$_3$	0.5084	0.5087	0.5093	0.5103	0.5120	
AgNO$_3$	0.4648	0.4652	0.4664	0.4682		
NH$_4$NO$_3$				0.4870		

$$\kappa = \alpha(u_+ + u_-)\mathscr{F}\tilde{c} \tag{8.24}$$

Table 8.5   Equivalent Conductivity in Water at 25°
(unit: cm^{2}S equiv^{-1})

Electrolyte	Λ°	Concentration/(eq/liter)		
		0.001	0.01	0.1
KCl	149.86	146.95	141.27	128.96
NaCl	126.45	123.74	118.51	106.74
HCl	426.16	421.36	412.00	391.32
AgNO$_3$	133.36	130.51	124.76	109.14
KNO$_3$	144.96	141.84	132.82	120.40
NH$_4$Cl	149.7	—	141.28	128.75
LiCl	115.03	112.40	107.32	95.86

$$\Lambda = \frac{\kappa}{\tilde{c}} \tag{8.25}$$

$$\Lambda = \alpha(u_+ + u_-)\mathscr{F} \tag{8.26}$$

Table 8.6 Limiting Ionic Conductivity ($\lambda°$) in Water (cm^2 S/equiv.)

Cations	18°C	25°C	Anions	18°C	25°C
H^+	315	349.8	OH^-	174	197.6
Li^+	32.55	38.69	F^-	47.6	55.4
Na^+	42.6	50.11	Cl^-	66.3	76.34
K^+	63.65	73.50	Br^-	68.2	78.14
Rb^+	66.3	77.8	I^-	66.8	76.97
Cs^+	66.8	77.3	CN^-	—	82
NH_4^+	63.6	73.4	NO_3^-	62.6	71.44
Ag^+	53.25	61.92	ClO_4^-	59.1	67.4
Tl^+	64.8	74.7	$(1/2)SO_4^{2-}$	68.7	80
$(1/2)Mg^{2+}$	44.6	53.06	$(1/3)Fe(CN)_6^{3-}$	—	99.1
$(1/2)Ca^{2+}$	50.4	59.50	$(1/4)Fe(CN)_6^{4-}$	—	111.
$(1/2)Sr^{2+}$	50.6	59.46	Formate$^-$	48	54.6
$(1/2)Zn^{2+}$	45.0	52.8	Acetate$^-$	35	40.9
$(1/2)Cd^{2+}$	45.1	54.	Chloroacetate$^-$	—	39.8
$(1/2)Pb^{2+}$	60.5	70.	Dichloroacetate$^-$	—	38
$(1/2)Mn^{2+}$	44.5	53.5	Trichloroacetate$^-$	—	35
$(1/3)Al^{3+}$	—	63.	n-Propionate$^-$	—	35.8
$(1/3)Co(NH_3)_6^{3+}$	—	99.2	Benzoate$^-$	—	32.3
$N(CH_3)_4^+$	—	44.92	Picrate$^-$	25.14	31.39
$N(C_2H_5)_4^+$	—	32.66			
$N(C_3H_7)_4^+$	—	23			

Source: G. Milazzo, *Electrochemistry,* 1963: Amsterdam, Elsevier Publishing Co.

$$\Lambda = \lambda_+ + \lambda_- \tag{8.28}$$

$$\lambda_i = u_i\mathscr{F}; \quad \lambda_i = t_i\Lambda \tag{8.29}$$

$$D_i = \frac{u_i RT}{\mathscr{F}|z_i|} \tag{8.30}$$

$$\overline{D}_{MX} = \frac{2D_{M^+}D_{X^-}}{D_{M^+} + D_{X^-}} \tag{8.31}$$

$$\alpha' = \frac{\Lambda}{\Lambda°} \tag{8.33}$$

$$\Delta_{rxn}G = -n\mathscr{F}\mathscr{E} \tag{8.38}$$

$$Q = \prod_i a_i^{\nu_i} \tag{8.39}$$

$$\mathscr{E} = \mathscr{E}^{\ominus} - \frac{RT}{n\mathscr{F}} \ln Q \tag{8.40}$$

$$\mathcal{E}^{\ominus} \equiv -\frac{\Delta_{rxn} G^{\ominus}}{n\mathcal{F}} \qquad (8.41)$$

$$\mathcal{E}^{\ominus}_{cell} = \mathcal{E}^{\ominus}_{right} - \mathcal{E}^{\ominus}_{left} \qquad (8.42)$$

Table 8.7 Standard Electrode Potentials in Water at 25°C, $P^{\ominus} = 1$ atm.*

Electrode	Electrode Reaction (Acid Solution)	$\mathcal{E}^{\ominus}$/Volts
$Li^+ \mid Li$	$Li^+ + e = Li$	-3.045
$K^+ \mid K$	$K^+ + e = K$	-2.925
$Ba^{2+} \mid Ba$	$Ba^{2+} + 2e = Ba$	-2.906
$Ca^{2+} \mid Ca$	$Ca^{2+} + 2e = Ca$	-2.866
$Na^+ \mid Na$	$Na^+ + e = Na$	-2.714
$Zn^{2+} \mid Zn$	$Zn^{2+} + 2e = Zn$	-0.7628
$Fe^{2+} \mid Fe$	$Fe^{2+} + 2e = Fe$	-0.4402
$Cd^{2+} \mid Cd$	$Cd^{2+} + 2e = Cd$	-0.4029
$SO_4^{2-} \mid PbSO_4 \mid Pb$	$PbSO_4 + 2e = Pb + SO_4^{2-}$	-0.3546
$I^- \mid AgI \mid Ag$	$AgI + e = Ag + I^-$	-0.1522
$Sn^{2+} \mid Sn$	$Sn^{2+} + 2e = Sn$	-0.136
$Pb^{2+} \mid Pb$	$Pb^{2+} + 2e = Pb$	-0.126
$Fe^{3+} \mid Fe$	$Fe^{3+} + 3e = Fe$	-0.036
$D^+ \mid D_2, Pt$	$2D^+ + 2e = D_2$	-0.0034
$H^+ \mid H_2, Pt$	$2H^+ + 2e = H_2$	(zero by convention)
$Br^- \mid AgBr \mid Ag$	$AgBr + e = Ag + Br^-$	0.0711
$Sn^{4+}, Sn^{2+} \mid Pt$	$Sn^{4+} + 2e = Sn^{2+}$	0.15
$Cu^{2+}, Cu^+ \mid Pt$	$Cu^{2+} + e = Cu^+$	0.153
$Cl^- \mid AgCl \mid Ag$	$AgCl + e = Ag + Cl^-$	0.2225
$Cl^- \mid Hg_2Cl_2 \mid Hg$	$Hg_2Cl_2 + 2e = 2Hg + 2Cl^-$	0.2680
$Cu^{2+} \mid Cu$	$Cu^{2+} + 2e = Cu$	0.337
$I^- \mid I_2 \mid Pt$	$I_2 + 2e = 2I^-$	0.5355
$Ag^+ \mid Ag$	$Ag^+ + e = Ag$	0.7991
$Hg^{2+} \mid Hg$	$Hg^{2+} + 2e = Hg$	0.854
$Hg^+ \mid Hg$	$Hg^+ + e = Hg$	0.92
$Br^- \mid Br_2 \mid Pt$	$Br_2 + 2e = 2Br^-$	1.0652
$Mn^{2+}, H^+ \mid MnO_2 \mid Pt$	$MnO_2 + 4H^+ + 2e = Mn^{2+} + 2H_2O$	1.208
$Cr^{3+}, Cr_2O_7^{2-}, H^+ \mid Pt$	$Cr_2O_7^{2-} + 14H^+ + 6e = 2Cr^{3+} + 7H_2O$	1.33
$Cl^- \mid Cl_2, Pt$	$Cl_2 + 2e = 2Cl^-$	1.3595
	(Basic solutions)	
$OH^- \mid Ca(OH)_2 \mid Ca \mid Pt$	$Ca(OH)_2 + 2e = 2OH^- + Ca$	-3.02
$ZnO_2^{2-}, OH^- \mid Zn$	$Zn(OH)_4^{2-} + 2e = Zn + 4OH^-$	-1.215
$OH^- \mid H_2 \mid Pt$	$2H_2O + 2e = H_2 + 2OH^-$	-0.82806
$CO_3^{2-} \mid PbCO_3 \mid Pb$	$PbCO_3 + 2e = Pb + CO_3^{2-}$	-0.509
$OH^- \mid HgO \mid Hg$	$HgO + H_2O + 2e = Hg + 2OH^-$	0.097

*To convert $P^{\ominus} = 0.1$ MPa $= 1$ bar, add $\Delta\mathcal{E}^{\ominus} = (0.3382$ mV$)(\Delta\nu_g/n)$ to the numbers on the table. $\Delta\nu_g$ is the change of moles of *gas* for the cell reaction [half cell opposed to $H^+ \mid H_2(g)$], and n is the number of electrons transferred.

$$\Delta_{rxn} S^{\ominus} = n \mathcal{F} \frac{d\mathcal{E}^{\ominus}}{dT} \tag{8.43}$$

$$\Delta_{rxn} H^{\ominus} = n \mathcal{F} \left(T \frac{d\mathcal{E}^{\ominus}}{dT} - \mathcal{E}^{\ominus} \right) \tag{8.44}$$

$$\mathcal{E} = -\frac{RT}{\mathcal{F}} \ln \left(\frac{a_{1+}a_{1-}}{a_{2+}a_{2-}} \right)^{t_-}$$

cation reversible
$$\mathcal{E} = -2t_- \frac{RT}{\mathcal{F}} \ln \frac{m_1 \gamma_{\pm 1}}{m_2 \gamma_{\pm 2}} \tag{8.51}$$

anion reversible
$$\mathcal{E} = 2t_+ \frac{RT}{\mathcal{F}} \ln \frac{m_1 \gamma_{\pm 1}}{m_2 \gamma_{\pm 2}} \tag{8.52}$$

Henderson
$$\mathcal{E}_J = \frac{RT}{\mathcal{F}} \left[\frac{c_1(\lambda_+ - \lambda_-) - c_2(\lambda'_+ - \lambda'_-)}{c_1 \Lambda - c_2 \Lambda'} \right] \ln \left[\frac{c_1 \Lambda}{c_2 \Lambda'} \right] \tag{8.54}$$

Donnan effect
$$\Pi = RT[c_P + z^2 c_P^2 / 4c_s] \tag{8.61}$$

CHAPTER 9

$$z = \sqrt{2} \, \bar{v} \pi \sigma^2 n^* \tag{9.8}$$

$$\lambda = \frac{1}{\sqrt{2} \, \pi \sigma^2 n^*} \tag{9.9b}$$

$$\langle v_{AB} \rangle = \left[\frac{8k_b T}{\pi \mu} \right]^{1/2} = \left[\frac{8RT}{\pi L \mu} \right]^{1/2} \tag{9.11}$$

$$z_{A:B} = \pi \sigma_{AB}^2 \langle v_{AB} \rangle n_B^* \tag{9.13a}$$

and the frequency of collisions by each B molecule with A molecule is:

$$z_{B:A} = \pi \sigma_{AB}^2 \langle v_{AB} \rangle n_A^* \tag{9.13b}$$

$$Z_{AA} = \frac{\pi \bar{v}_A \sigma_A^2 (n_A^*)^2}{\sqrt{2}} \tag{9.14}$$

$$Z_{AB} = \pi \sigma_{AB}^2 \langle v_{AB} \rangle n_A^* n_B^* \tag{9.15}$$

$$C(n, p) = \frac{n!}{p! \, (n - p)!} \tag{9.16}$$

$$\langle m^2 \rangle = n; \qquad \langle m^2 \rangle^{1/2} = \sqrt{n} \tag{9.17}$$

$$W(n, m) = \frac{2}{\sqrt{2\pi n}} e^{-m^2/2n} \mathcal{P}_{n, m} \tag{9.18}$$

$$W(x, n) \, dx = \frac{1}{\sqrt{2\pi n \lambda^2}} e^{-x^2/2n\lambda^2} \, dx \tag{9.19}$$

$$W(x, t)dx = \frac{1}{\sqrt{4\pi Dt}} e^{-x^2/2Dt} \, dx \tag{9.20}$$

$$D = \frac{z\lambda^2}{2} \tag{9.21}$$

$$\delta x_{rms} = \sqrt{2Dt} \tag{9.22}$$

2D
$$W(r, t) \, dr = \frac{1}{2Dt} e^{-r^2/4Dt} r \, dr \tag{9.23}$$

3D
$$W(r, t) \, dr = \frac{4\pi}{(4\pi Dt)^{3/2}} e^{-r^2/4Dt} r^2 dr \tag{9.24}$$

$$W_r(n, m, m_1) = W(n, m) + W(n, 2m_1 - m)$$
$$W_a(n, m, m_1) = W(n, m) - W(n, 2m_1 - m) \tag{9.25}$$

$$J = \frac{1}{A}\frac{dn}{dt} = -D\frac{dc}{dx} \tag{9.27}$$

$$\frac{\partial c}{\partial t} = D\frac{\partial^2 c}{\partial x^2} \tag{9.29}$$

$$c(x, t) = \frac{c_0 \, \Delta x}{2\sqrt{\pi Dt}} e^{-x^2/4Dt} \tag{9.32}$$

$$D = \tfrac{1}{2}\bar{v}\,\lambda \tag{9.33}$$

$$\xi \equiv x/\sqrt{4Dt} \tag{9.34}$$

Table 9.1   Diffusion Constants

Solute	Solvent	$t/°C$	$10^5 \, D/(cm^2 \, s^{-1})$
Water	(Self-diffusion)	25	2.26
Iodine	n-Hexane	25	4.05
Iodine	CCl_4	25	2.13
Argon	CCl_4	25	3.63
CCl_4	n-Heptane	25	3.17
Glycine	Water	25	1.05
Glycine	Water	20	0.93
Dextrose	Water	25	0.67
Sucrose	Water	20	0.4586
Ribonuclease (a)*	Water	20	0.1068
Myosin (b)*	Water	20	0.0105
Rabbit papilloma virus (c)*	Water	20	0.0059

* Molecular weights: (a) 13 683 g, (b) 4.4×10^5 g, (c) 4.7×10^7 g.

$$F = \eta A \frac{dv}{dx} \tag{9.36}$$

Table 9.2   Viscosity of Some Fluids (units: p = g cm^{-1} s^{-1}; temperature in parentheses)

Liquids	η/mp				
Water	17.921(0°C)	10.050(20°C)	8.937(25°C)	8.007(30°C)	5.494(50°C)
Acetic acid	12.22(20°C)	10.396(30°C)	7.956(50°C)		
Acetone	4.013(0°C)	3.311(20°C)	2.561(50°C)		
Cyclohexane	10.3(17°C)	8.6(27°C)	7.5(35°C)		
n-Heptane	5.236(0°C)	4.163(20°C)	3.410(40°C)		
Ethylene glycol		173.3(25°C)			
Glycerol	42 200(2.8°C)	10 690(20°C)			

Gases	η/μp		
Nitrogen	129.5 (200 K)	178.6 (300 K)	401.1 (1000 K)
Oxygen	147.6 (200 K)	207.1 (300 K)	472.0 (1000 K)
Argon	159.4 (200 K)	227.0 (300 K)	530.2 (1000 K)
CO_2(20°C)	148 (1 atm)	156 (20 atm)	166 (40 atm)

$$\frac{\Delta V}{\Delta t} = \frac{\pi r^4 \, \Delta P}{8 \eta l} \tag{9.37}$$

$$\frac{\Delta V}{\Delta t} = \frac{\pi r^4}{16 \eta l} \left(\frac{P_i^2 - P_f^2}{P_0} \right) \tag{9.38}$$

$$v_s = \frac{2r^2(\rho - \rho_0)g}{9\eta} \tag{9.40}$$

Stokes-Einstein law
$$D = \frac{k_b T}{6\pi \eta r} \tag{9.42}$$

$$\eta = 0.5 \rho \overline{v} \lambda \tag{9.46}$$

Sutherland
$$\eta = \frac{k_s \sqrt{T}}{1 + \dfrac{S}{T}} \tag{9.48}$$

$$k_s = \frac{\sqrt{RM}}{\pi^{3/2} \sigma^2 L} \tag{9.49}$$

$$n_{sp} \equiv \frac{\eta}{\eta_0} - 1 = \frac{\eta - \eta_0}{\eta_0} \tag{9.50}$$

$$[\eta] \equiv \lim_{c_m \to 0} \left(\frac{\eta_{sp}}{c_m} \right) \tag{9.51}$$

$$[\eta] = KM^{\alpha} \tag{9.54}$$

Table 9.3 Intrinsic Viscosity Parameters

Polymer	Solvent	$100K/(\text{cm}^3/\text{g})$	α	$\Gamma(1 + \alpha)$	$t/(°C)$
Polystyrene	Cyclohexane	8.1	0.50	0.886	34°
Polystyrene	Benzene	0.95	0.74	0.917	25°
Polystyrene	Toluene	3.7	0.62	0.896	25°
Polyisobutylene	Benzene	8.3	0.50	0.886	24°
Polyisobutylene	Cyclohexane	2.6	0.70	0.909	30°

$$M_v/M_n = [(1 + \alpha)\,\Gamma\,(1 + \alpha)]^{1/\alpha} \tag{9.56}$$

$$\frac{dx}{dt} = \frac{MDg}{RT}(1 - \bar{v}_s\rho_0) \tag{9.57}$$

$$\frac{dx}{dt} = \frac{MD\omega^2 x(1 - \bar{v}_s\rho_0)}{RT} \tag{9.58}$$

$$s \equiv \frac{(dx/dt)}{\omega^2 x} \tag{9.60}$$

$$M = \frac{RTs}{D(1 - \rho_0\bar{v})} \tag{9.61}$$

Table 9.4 Properties of Biological Macromolecules (20°C, in water)

Molecule	M (g/mol)	$[\eta]$ (cm^3/g)	$10^7 D$ (cm^2/s)	v_s (cm^3/g)	$10^{13} s$ (sec)
Ribonuclease	13 683	2.30	11.9	0.728	1.64
Lysozyme	14 400	—	11.2	0.703	1.91
Serum albumin	66×10^3	3.7	5.94	0.734	4.31
Hemoglobin	68×10^3	3.6	6.9	0.749	4.83
Myosin	570×10^3	217	1.0	0.728	6.4
Catalase	250×10^3	3.9	4.1	0.73	11.3
Bushy stunt virus	10.7×10^6	3.4	1.15	0.74	132
Tobacco mosaic virus	50×10^6	36.7	0.3	0.73	170

$$\frac{\ln\left(\dfrac{c_2}{c_1}\right)}{x_2^2 - x_1^2} = \frac{M(1 - \rho_0\bar{v}_s)\omega^2}{2RT} \tag{9.62}$$

CHAPTER 10

$$v = \frac{1}{\nu_i}\frac{dC_i}{dt} \tag{10.1}$$

$$\ln\frac{C_0}{C} = kt \tag{10.5a}$$

$$\lambda - \lambda_\infty = (\lambda_0 - \lambda_\infty)e^{-kt} \qquad \textbf{(10.6)}$$

$$\frac{1}{C} = \frac{1}{C_0} + kt \qquad \textbf{(10.8)}$$

$$\frac{1}{\lambda - \lambda_\infty} = \frac{1}{\lambda_0 - \lambda_\infty} + \frac{C_0 kt}{\lambda_0 - \lambda_\infty} \qquad \textbf{(10.10)}$$

$$k = \frac{1}{t(b - a)} \ln\left[\frac{a(b - x)}{b(a - x)}\right] \qquad \textbf{(10.11b)}$$

$$\frac{1}{C^{n-1}} = \frac{1}{C_0^{n-1}} + (n - 1)kt \qquad \textbf{(10.12)}$$

$$\ln k = \ln A - \frac{E_a}{RT}; \qquad k = Ae^{-E_a/RT} \qquad \textbf{(10.18)}$$

Table 10.1   Arrhenius Factors for Various Reactions

First Order	$A/(\text{s}^{-1})$	$E_a/(\text{kJ/mol})$
cis → *trans*-1, 2-dideuterocyclopropane	2.5×10^{16}	272
$CH_3NC \rightarrow CH_3CN$	4×10^{13}	160
$C_2H_5I \rightarrow C_2H_4 + HI$	2.5×10^{13}	209
$C_2H_6 \rightarrow 2CH_3$	2.5×10^{17}	384
$N_2O_5 \rightarrow NO_2 + NO_3$	6.3×10^{14}	88
$CH_3CO \rightarrow CH_3 + CO$	1×10^{15}	43
Second Order	$A/(\text{cm}^3\text{mol}^{-1}\text{s}^{-1})$	$E_a/(\text{kJ/mol})$
$NO + O_3 \rightarrow NO_2 + O_2$	7.9×10^{11}	10.5
$2NOCl \rightarrow 2NO + Cl_2$	1×10^{13}	103.6
$NO + Cl_2 \rightarrow NOCl + Cl$	4.0×10^{12}	84.9
$CH_3 + CH_3 \rightarrow C_2H_6$	2×10^{13}	~0
$OH + H_2 \rightarrow H_2O + H$	8×10^{13}	42
$Cl + H_2 \rightarrow HCl + H$	8×10^{13}	23
$O_3 + C_3H_8 \rightarrow C_3H_7O + HO_2$	10^9	51
$2NOBr \rightarrow 2NO + Br_2$	4.15×10^{13}	58.1
Third Order	$A/(\text{cm}^6\text{mol}^{-2}\text{s}^{-1})$	$E_a/(\text{kJ/mol})$
$I + I + Ar \rightarrow I_2 + Ar$	6.3×10^{15}	5.4
$H + H + H_2 \rightarrow H_2 + H_2$	1×10^{16}	~0
$O + O + O_2 \rightarrow O_2 + O_2$	1×10^{15}	~0
$O + O_2 + Ar \rightarrow O_3 + Ar$	3.2×10^{12}	−9.6
$2NO + O_2 \rightarrow 2NO_2$	1.05×10^9	−4.6
$2NO + Cl_2 \rightarrow 2NOCl$	1.7×10^{10}	15
$2NO + Br_2 \rightarrow 2NOBr$	3.2×10^9	~0

SCT

$$k = pLS_{AB} \sqrt{\frac{8RT}{\pi(L\mu)}} \exp\left(-\frac{E_{\min}}{RT}\right) \qquad \textbf{(10.23a)}$$

Eyring equation

$$k = \frac{k_b T}{hC^{\ominus}} e^{\Delta S^{\ddagger}/R} e^{-\Delta H^{\ddagger}/RT} \qquad \textbf{(10.25)}$$

Table 10.2 Arrhenius Factors for Radical Reactions

	$\log_{10} A/(cm^3\ mol^{-1}s^{-1})$	$E_a/(kJ/mol)$
$Cl + H_2 \rightarrow HCl + H$	13.9	23
$Br + H_2 \rightarrow HBr + H$	13.9	72
$I + H_2 \rightarrow HI + H$	14.1	140
$H + HCl \rightarrow H_2 + Cl$	13.6	19
$H + HBr \rightarrow H_2 + Br$	13.2	4
$2CH_3 \rightarrow C_2H_6$	13.3	~0
$2C_2H_5 \rightarrow C_4H_{10}$	13.2	~0
$Cl + COCl + COCl_2$	14.6	3.3
Termolecular recombinations (A in $cm^6\ mol^{-2}s^{-1}$):		
$H + H + H_2 \rightarrow H_2 + H_2$	16.0	~0
$O + O + O_2 \rightarrow O_2 + O_2$	15.0	~0
$O + O_2 + Ar \rightarrow O_3 + Ar$	12.5	−9.6
$I + I + Ar \rightarrow I_2 + Ar$	15.8	−5.4

Source: C. Walling, *Free Radicals in Solution*, 1975: New York, John Wiley & Sons.

kinetic chain length

$$\nu = \frac{\text{velocity of propagation}}{\text{velocity of initiation}} \qquad (10.44)$$

quantum yield

$$\phi = \frac{k_1 k_3}{k_3 + k_2[M]} \qquad (10.46)$$

$$b = k_a/k_d$$

$$\theta = \frac{bP}{1 + bP} \qquad (10.53)$$

$$\theta_i = \frac{b_i P_i}{1 + \Sigma_n b_n P_n} \qquad (10.57)$$

$$k = 4\pi L(D_A + D_B)\sigma_{AB} \qquad (10.66)$$

$$k = \left(\frac{2RT}{3\eta}\right)\frac{(r_A + r_B)^2}{r_A r_B} \qquad (10.67)$$

$$k = \left(\frac{k_b T}{hC^\ominus}\right)\left(\frac{\gamma_A \gamma_B}{\gamma^\ddagger}\right) e^{-\Delta s^\ddagger} e^{-\Delta H^\ddagger/RT} \qquad (10.68)$$

Table 10.3 Rate Constants in Aqueous Solution (25°)

Reaction	k(forward) $(liter\ mol^{-1}s^{-1})$	k(reverse) (s^{-1})
$H^+ + OH^- = H_2O$	1.4×10^{11}	2.5×10^{-5}
$D^+ + OD^- = D_2O$	8.4×10^{10}	2.5×10^{-6}
$H^+ + CH_3COO^- = CH_3COOH$	4.5×10^{10}	7.8×10^5
$H^+ + C_6H_5COO^- = C_6H_5COOH$	3.5×10^{10}	2.2×10^6
$H^+ + NH_3 = NH_4^+$	4.3×10^{10}	24.6
$OH^- + NH_4^+ = NH_3 + H_2O$	3.4×10^{10}	6×10^5

Source: F. Daniels and R. A. Alberty, *Physical Chemistry*, 3d ed., 1966: New York, John Wiley & Sons Inc.

CHAPTER 11 ▭▭▭▭▭▭▭▭▭▭▭▭▭▭▭▭▭▭▭▭▭

Bohr
$$r = \frac{N^2}{Z} a_0 \tag{11.3}$$

$$E = -\frac{Z^2}{N^2} \frac{e^4 m_e}{2\hbar^2 (4\pi\epsilon_0)^2} = -\frac{1}{2} \frac{Z^2}{N^2} E_h \tag{11.5}$$

$$E_n = \frac{n^2 h^2}{8ma^2} \tag{11.25}$$

$$\psi_n = \sqrt{\frac{2}{a}} \sin\left(\frac{n\pi x}{a}\right) \tag{11.26}$$

$$E = (n_x^2 + n_x^2 + n_z^2)\frac{h^2}{8ma^2} \tag{11.29}$$

$$\Psi(x, y, z) = \left(\frac{2}{a}\right)^{3/2} \sin\frac{n_x \pi x}{a} \sin\frac{n_y \pi y}{a} \sin\frac{n_z \pi z}{a} \tag{11.30}$$

$$\alpha = \left(\frac{\hbar^2}{k\mu}\right)^{1/4} \tag{11.33}$$

$$\psi_n = A_n H_n(y) e^{-y^2/2} \tag{11.35}$$

$$\nu_0 = \frac{1}{2\pi} \sqrt{\frac{k}{\mu}} \tag{11.36}$$

$$E_n = (n + \tfrac{1}{2})h\,\nu_0 \tag{11.37}$$

Table 11.2   Hermite Polynomials[*]

n	$H_n(y)$
0	1
1	$2y$
2	$4y^2 - 2$
3	$8y^3 - 12y$
4	$16y^4 - 48y^2 + 12$
5	$32y^5 - 160y^3 + 120y$
6	$64y^6 - 480y^4 + 720y^2 - 120$
⋮	
12	$4096y^{12} - 135\,168y^{10} + 1\,520\,640y^8 - 7\,096\,320y^6$ $+ 13\,305\,600y^4 - 7\,983\,360y^2 + 665\,280$

[*]Recursion formula: $H_{n+1} = 2y\,H_n - 2n\,H_{n-1}$

$$\hat{L}_x = \frac{1}{i}\left(y\frac{\partial}{\partial z} - z\frac{\partial}{\partial y}\right); \quad \hat{L}_y = \frac{1}{i}\left(z\frac{\partial}{\partial x} - x\frac{\partial}{\partial z}\right); \quad \hat{L}_z = \frac{1}{i}\left(x\frac{\partial}{\partial y} - y\frac{\partial}{\partial z}\right) \tag{11.40}$$

$$[\hat{L}_x, \hat{L}_y] = i\hat{L}_z; \qquad [[\hat{L}_y, \hat{L}_z] = i\hat{L}_x; \qquad [\hat{L}_z, \hat{L}_x] = i\hat{L}_y \tag{11.41}$$

$$\hat{L}_x = i\left(\sin\phi\,\frac{\partial}{\partial\theta} + \cot\theta\cos\phi\,\frac{\partial}{\partial\phi}\right) \tag{11.42a}$$

$$\hat{L}_y = -i\left(\cos\phi\,\frac{\partial}{\partial\theta} - \cot\theta\sin\phi\,\frac{\partial}{\partial\phi}\right) \tag{11.42b}$$

$$\hat{L}_z = -i\,\frac{\partial}{\partial\phi} \tag{11.42c}$$

$$\hat{L}^2 = -\left[\frac{1}{\sin\theta}\left(\frac{\partial}{\partial\theta}\left(\sin\theta\frac{\partial}{\partial\theta}\right)\right) + \frac{1}{\sin^2\theta}\frac{\partial^2}{\partial\phi^2}\right] \tag{11.43b}$$

$$E_m = m^2\frac{\hbar^2}{2I} = m^2\frac{h^2}{8\pi^2 I} \tag{11.44}$$

$$\psi_m = \frac{1}{\sqrt{2\pi}}e^{im\phi} \quad (m = 0,\ \pm 1,\ \pm 2\ \dots\ ) \tag{11.45}$$

$$\int \psi_{l,m}^*\psi_{l',m'}\,d\tau \begin{cases} = 1 & \text{if } l = l' \quad\text{and}\quad m = m' \\ = 0 & \text{if } l \neq l' \quad\text{or}\quad m \neq m' \end{cases} \tag{11.54}$$

$$\hat{L}^2\psi_{l,m} = l(l+1)\psi_{l,m}; \qquad \hat{L}_z\psi_{l,m} = m\psi_{l,m} \tag{11.55}$$

$$\psi_{l,m} = Y_{l,m}(\theta,\ \phi) = A P_l^{|m|}(\theta)\,e^{im\phi} \tag{11.57}$$

See Table 11.7 on page 45.

$$E_l = l(l+1)\frac{\hbar^2}{2I} = l(l+1)\frac{h^2}{8\pi^2 I} \tag{11.59}$$

$$\nu = \frac{\Delta E}{h} \tag{11.60a}$$

$$\tilde{\nu} = \frac{\Delta E}{hc} \tag{11.60b}$$

$$\mu_x = qx = qr\sin\theta\cos\phi$$
$$\mu_y = qy = qr\sin\theta\sin\phi \tag{11.61}$$
$$\mu_2 = qz = qr\cos\theta$$

$$I_x \equiv \int \psi_i^* x\psi_j\,d\tau; \qquad I_y \equiv \int \psi_i^* y\psi_j\,d\tau; \qquad I_z \equiv \int \psi_i^* z\psi_j\,d\tau \tag{11.62}$$

$$\text{intensity} \propto I_x^2 + I_y^2 + I_z^2 \tag{11.63}$$

See Table 11.8 on page 45.

Table 11.7   Spherical Harmonics and Associated Legendre Polynomials

$$Y_{l,m} = AP_l^{|m|}(\theta)\, e^{im\phi}$$

l	$|m|$	A	$P_l^{|m|}(\theta)$
0	0	$\sqrt{\dfrac{1}{4\pi}}$	1
1	0	$\sqrt{\dfrac{3}{4\pi}}$	$\cos\theta$
1	1	$\sqrt{\dfrac{3}{8\pi}}$	$\sin\theta$
2	0	$\sqrt{\dfrac{5}{16\pi}}$	$3\cos^2\theta - 1$
2	1	$\sqrt{\dfrac{15}{8\pi}}$	$\sin\theta\cos\theta$
2	2	$\sqrt{\dfrac{15}{32\pi}}$	$\sin^2\theta$
3	0	$\sqrt{\dfrac{7}{16\pi}}$	$\cos\theta\,(5\cos^2\theta - 3)$
3	1	$\sqrt{\dfrac{21}{64\pi}}$	$\sin\theta\,(5\cos^2\theta - 1)$
3	2	$\sqrt{\dfrac{105}{32\pi}}$	$\sin^2\theta\cos\theta$
3	3	$\sqrt{\dfrac{35}{64\pi}}$	$\sin^3\theta$

Table 11.8   Regions of Spectroscopy

Region	Phenomena	Approximate Range		
		Energy	Wavelength	Frequency
Radiofrequency	Nuclear spin	—	300 m–30 cm	1 MHz – 1 GHz
Microwave	Electron spin, molecular rotation	—	30 cm–1 mm	1 GHz–300 GHz
Optical-infrared	Molecular vibration	—	1 mm–750 nm	10–13333 cm^{-1}
Optical-visible	Electrons	1.6 eV–3 eV	750 nm–400 nm	13333–25000 cm^{-1}
Optical-ultraviolet	Electrons	3 eV–100 eV	400 nm–10 nm	25000–10^6 cm^{-1}
X-ray	Inner electrons	100 eV up	10 nm down	—
γ-ray	Nuclear binding	ca. 100 keV	—	—

CHAPTER 12

$$E_n = -\frac{Z^2 E_h}{2}\frac{1}{n^2} \quad \textbf{(12.4)} \qquad R_{nl} = AL_{nl}(Z\sigma)Z^l\sigma^l e^{-Z\sigma/n} \qquad \textbf{(12.5)}$$

$$\psi_{nlm} = R_{nl}(r)\, Y_{lm}(\theta, \phi) \qquad \textbf{(12.6)}$$

Table 12.1 Radial Wave Functions for the One-Electron Atom

					$R_{nl} = AL_{nl}(Z\sigma)Z^l\sigma^l e^{-Z\sigma/n}$
n	l	$L_{nl}(\xi)$	n	l	$L_{nl}(\xi)$
1	0	1	4	0	$(192 - 144\xi + 24\xi^2 - \xi^3)/48$
2	0	$2 - \xi$	4	1	$(80 - 20\xi + \xi^2)/8$
2	1	1	4	2	$6 - \xi/2$
3	0	$3 - 2\xi + 2\xi^2/9$	4	3	1
3	1	$4 - 2\xi/3$	5	0	$(9375 - 7500\xi + 1500\xi^2 - 100\xi^3 + 2\xi^4)/1875$
3	2	1	5	1	$20 - 6\xi + 12\xi^2/25 - 4\xi^3/375$
			5	2	$21 - 14\xi/5 + 2\xi^2/25$
			5	3	$8 - 2\xi/5$

radial distribution function $\qquad f_r = R^2 r^2\, dr \qquad\qquad$ **(12.7)**

$$E_{var} = -(Z - \tfrac{5}{16})^2 E_h \qquad \textbf{(12.15)}$$

Table 12.6 Allowed Terms of Equivalent Electrons

Configurations	Terms
p^2, p^4	$^1S, {}^1D, {}^3P$
p^3	$^2P, {}^2D, {}^4S$
p, p^5	^2P
d, d^9	^2D
d^2, d^8	$^1S, {}^1D, {}^1G, {}^3P, {}^3F$
d^3, d^7	$^2P, {}^2D$ (twice), $^2F, {}^2G, {}^2H, {}^4P, {}^4F$
d^4, d^6	^1S (twice), ^1D (twice), $^1F, {}^1G$ (twice), $^1I, {}^3P$ (twice), ^3D, ^3F (twice), $^3G, {}^5D$
d^5	$^2S, {}^2P, {}^2D$ (thrice), ^2F (twice), ^2G (twice), $^2H, {}^2I, {}^4P$, $^4D, {}^4F, {}^4G, {}^6S$

$$E_{SO} = \tfrac{1}{2}hcA[J(J + 1) - L(L + 1) - S(S + 1)] \qquad \textbf{(12.30)}$$

$$|\mu_J| = g_L\mu_B\sqrt{J(J + 1)}; \qquad \mu_z = g_L\mu_B M_J \qquad \textbf{(12.33)}$$

$$g_L = 1 + \frac{J(J + 1) + S(S + 1) - L(L + 1)}{2J(J + 1)} \qquad \textbf{(12.34)}$$

$$E_Z = -M_J g_L\mu_B B_0 \qquad \textbf{(12.35)}$$

ESR $\qquad\qquad \nu = \Delta E/h = g\mu_B B_0/h \qquad \textbf{(12.36)}$

NMR $\qquad\qquad \nu = \dfrac{g_N\mu_N B_0}{h} = \dfrac{\gamma}{2\pi}B_0 \qquad \textbf{(12.37)}$

$$\hat{H}_{hf} = hA\hat{\mathbf{I}} \cdot \hat{\mathbf{S}} \tag{12.39}$$

$$E_{hf}/h = \tfrac{1}{2}A[T(T + 1) - I(I + 1) - S(S + 1)] \tag{12.42}$$

CHAPTER 13

$$k_e \equiv \left(\frac{\partial^2 E_e}{\partial R^2}\right)_{R_e} \tag{13.11}$$

Table 13.1   Ground-State Molecular Constants for Some Diatomic Molecules

Molecule	Term Symbol	ω_e/cm^{-1}	$\omega_e x_e/\text{cm}^{-1}$	$\tilde{B}_e/\text{cm}^{-1}$	α_e/cm^{-1}	$R_e/\text{Å}$	D_0/eV
$^1\text{H}^1\text{H}$	$^1\Sigma_g^+$	4395.2	117.91	60.81	2.993	0.7417	4.4773
$^{12}\text{C}^{16}\text{O}$	$^1\Sigma^+$	2170.21	13.461	1.9314	0.01748	1.1282	11.108
$^1\text{H}^{35}\text{Cl}$	$^1\Sigma^+$	2989.74	52.05	10.5909	0.3019	1.27460	4.436
$^{35}\text{Cl}^{35}\text{Cl}$	$^1\Sigma_g^+$	564.9	4.0	0.2438	0.0017	1.988	2.475
$^{14}\text{N}^{14}\text{N}$	$^1\Sigma_g^+$	2359.61	14.456	2.010	0.0187	1.094	9.756
$^1\text{H}^{79}\text{Br}$	$^1\Sigma^+$	2649.67	45.21	8.473	0.226	1.414	3.775
$^1\text{H}^{127}\text{I}$	$^1\Sigma^+$	2309.5	39.73	6.551	0.183	1.604	3.053
$^{127}\text{I}^{35}\text{Cl}$	$^1\Sigma^+$	384.18	1.465	0.1141619	0.00053	2.32069	2.152
$^{127}\text{I}^{79}\text{Br}$	$^1\Sigma^+$	268.4	0.78	—	—	—	1.817
$^1\text{H}^{19}\text{F}$	$^1\Sigma^+$	4138.52	90.069	20.939	0.770	0.9171	5.86
$^2\text{D}^{19}\text{F}$	$^1\Sigma^+$	2998.25	45.71	11.007	0.293	0.9170	—

Morse potential

$$E_e(R) = D_e[1 - e^{-\beta(R-R_e)}]^2 \tag{13.13}$$

$$\beta = 2\pi c\omega_e \sqrt{\frac{\mu}{2D_e}} \tag{13.14}$$

$$E_n^{\text{vib}} = hc[(n + \tfrac{1}{2})\omega_e - (n + \tfrac{1}{2})^2\omega_e x_e + (n + \tfrac{1}{2})^3\omega_e y_e] \cdots \tag{13.15}$$

Table 13.2   Nuclear Masses and Abundance

Nucleus	Mass/(g/mol)	% Natural Abundance	Nucleus	Mass/(g/mol)	% Natural Abundance
^1H	1.0078250	99.985	^{19}F	18.9984032	100
^2D	2.0141018	0.015	^{23}Na	22.989768	100
^6Li	6.0151214	7.5	^{32}S	31.972071	95.12
^7Li	7.0160030	92.5	^{34}S	33.967867	4.21
^{12}C	12 (exactly)	98.90	^{35}Cl	34.9688527	75.77
^{13}C	13.0033548	1.10	^{37}Cl	36.9659026	24.23
^{14}N	14.0030740	99.634	^{39}K	38.963707	93.2581
^{15}N	15.00010897	0.366	^{79}Br	78.918336	50.69
^{16}O	15.9949146	99.762	^{81}Br	80.916289	49.31
^{17}O	16.9991312	0.038	^{127}I	126.904473	100
^{18}O	17.9991603	0.200	^{133}Cs	132.905429	100

Source: IUPAC, "Quantities, Units and Symbols in Physical Chemistry," 1988: Oxford, England, Blackwell Scientific Publications.

$$\omega'_e = \omega_e \left[\frac{\mu_{AX}}{\mu_{AX'}}\right]^{1/2} \tag{13.18}$$

$$\omega'_e x'_e = (\omega_e x_e)\frac{\mu_{AX}}{\mu_{AX'}} \tag{13.19}$$

$$D_c = \frac{4B_e^3}{\omega_e^2 c^2} \tag{13.23}$$

$$E_J/h = J(J+1)B_n - J^2(J+1)^2 D_c; \quad \text{with } B_n = B_e - \left(n + \tfrac{1}{2}\right)\alpha_e \tag{13.24}$$

$$\frac{E_{n,J}}{hc} = \left(n + \tfrac{1}{2}\right)\omega_e - \left(n + \tfrac{1}{2}\right)^2 \omega_e x_e \tag{13.25}$$

$$+ J(J+1)B_e - J(J+1)\left(n + \tfrac{1}{2}\right)\alpha_e - J^2(J+1)^2 D_c$$

$$\sigma_g 1s,\ \sigma_u^* 1s,\ \sigma_g 2s,\ \sigma_u^* 2s,\ \pi_u 2p,\ \sigma_g 2p,\ \pi_g^* 2p,\ \sigma_u^* 2p \tag{13.46}$$

$$\tilde{\nu}_{v'-0} = \tilde{\nu}_{00} + v'(\omega'_e - \omega'_e x'_e) - (v')^2 \omega'_e x'_e \tag{13.50}$$

Table 13.7 Properties of Alkali Halide Diatomics

	ω_e/cm^{-1}	$R_e/\text{Å}$	D_e/eV	μ/debye
NaCl	364.6	2.3609	4.22	9.00
KCl	279.8	2.6668	4.37	10.48
KBr	219.17	2.8208	3.92	10.41
RbI	119.20	3.3152	3.57	12.1
LiF	910.34	1.5639	5.99	6.32

Table 13.8 Ionization Potentials and Electron Affinities

Atom	Z	IP/eV	EA/eV
H	1	13.595	0.7542
Li	3	5.390	0.620
C	6	11.256	1.268
O	8	13.614	1.462
F	9	17.42	3.399
Na	11	5.138	0.546
Cl	17	13.01	3.615
K	19	4.339	0.5012
Br	35	11.84	3.364
Rb	37	4.176	0.4860

Source: *J. Phys. Chem. Ref. Data*, **4**, 539 (1975).

$$\frac{\epsilon - 1}{\epsilon + 2}\frac{M}{\rho} = \frac{4\pi L}{3}\left(\alpha + \frac{\mu_0^2}{3(4\pi\epsilon_0)k_b T}\right) \tag{13.56}$$

$$E_{\text{Stark}} = -\frac{I\mu_0^2 \mathscr{E}^2}{\hbar^2}\left[\frac{J(J+1) - 3M^2}{J(J+1)(2J-1)(2J+3)}\right] \tag{13.58}$$

CHAPTER 14 ▨▨▨▨▨▨▨▨▨▨▨▨▨▨▨▨▨▨▨▨▨

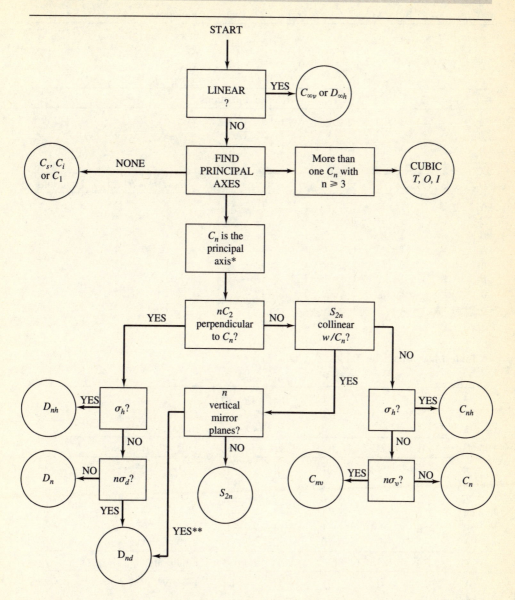

Figure 14.2   Flow diagram for determining point groups. The inversion operation is found in the groups $C_i = S_2$, C_{nh} (if n is even), D_{nh} (if n is even), D_{nd} (if n is odd), $D_{\infty h}$ (linear), S_6, and some of the cubic groups. The only cubic groups of much importance in chemistry are T_d, O_h, and I_h. Notes: *If there are three mutually perpendicular C_2 axes, choose the principal axis to the one that passes through the most (or the heaviest) atoms. ** If such is the case, there are n perpendicular $\hat{C}_2$ axes—try again to see them; building a model will certainly help. [J. Noggle, *J. Chem. Ed.,* **53**, 178 (1976).]

Table 14.1 Character Tables: C_{2v}, C_{2h}, D_{2h}

C_{2v}	$\hat{E}$	$\hat{C}_2(z)$	$\hat{\sigma}_v(xz)$	$\hat{\sigma}_v'(yz)$	Functions
$A_1\ (a_1)$	1	1	1	1	z, x^2, y^2, z^2
$A_2\ (a_2)$	1	1	-1	-1	xy
$B_1\ (b_1)$	1	-1	1	-1	x, xz
$B_2\ (b_2)$	1	-1	-1	1	y, yz

C_{2h}	$\hat{E}$	$\hat{C}_2$	$\hat{\imath}$	$\hat{\sigma}_h$	Functions
A_g	1	1	1	1	x^2, y^2, z^2, xy
B_g	1	-1	1	-1	xz, yz
A_u	1	1	-1	-1	z
B_u	1	-1	-1	1	x, y

D_{2h}	$\hat{E}$	$\hat{C}_2(z)$	$\hat{C}_2(y)$	$\hat{C}_2(x)$	$\hat{\imath}$	$\hat{\sigma}(xy)$	$\hat{\sigma}(xz)$	$\hat{\sigma}(yz)$	Functions
A_g	1	1	1	1	1	1	1	1	x^2, y^2, z^2
B_{1g}	1	1	-1	-1	1	1	-1	-1	xy
B_{2g}	1	-1	1	-1	1	-1	1	-1	xz
B_{3g}	1	-1	-1	1	1	-1	-1	1	yz
A_u	1	1	1	1	-1	-1	-1	-1	
B_{1u}	1	1	-1	-1	-1	-1	1	1	z
B_{2u}	1	-1	1	-1	-1	1	-1	1	y
B_{3u}	1	-1	-1	1	-1	1	1	-1	x

Table 14.2 Character Tables: C_{3v}, D_{3h}, T_d

C_{3v}	$\hat{E}$	$2\hat{C}_3$	$3\hat{\sigma}_v$	Functions[a]
A_1	1	1	1	$z, x^2 + y^2, z^2$
A_2	1	1	-1	
E	2	-1	0	$(x, y)(x^2 - y^2, xy)(xz, yz)$

D_{3h}	$\hat{E}$	$2\hat{C}_3$	$3\hat{C}_2$	$\hat{\sigma}_h$	$2\hat{S}_3$	$3\hat{\sigma}_v$	Functions[a]
A_1'	1	1	1	1	1	1	$z^2, x^2 + y^2$
A_2'	1	1	-1	1	1	-1	
E'	2	-1	0	2	-1	0	$(x, y)(x^2 + y^2, xy)$
A_1''	1	1	1	-1	-1	-1	
A_2''	1	1	-1	-1	-1	1	z
E''	2	-1	0	-2	1	0	(xz, yz)

T_d	$\hat{E}$	$8\hat{C}_3$	$3\hat{C}_2$	$6\hat{S}_4$	$6\hat{\sigma}_d$	Functions[a]
A_1	1	1	1	1	1	$x^2 + y^2 + z^2$
A_2	1	1	1	-1	-1	
E	2	-1	2	0	0	$(2z^2 - x^2 - y^2, x^2 - y^2)$
T_1	3	0	-1	1	-1	
T_2	3	0	-1	-1	1	$(x, y, z)(xy, xz, yz)$

[a] Groups in parenthesis are degenerate.

Table 14.3 Character Tables: D_{6h}, O_h

D_{6h}	$\hat{E}$	$2\hat{C}_6$	$2\hat{C}_3$	$\hat{C}_2$	$3\hat{C}_2'$	$3\hat{C}_2''$	$\hat{i}$	$2\hat{S}_3$	$2\hat{S}_6$	$\hat{\sigma}_h$	$3\hat{\sigma}_d$	$3\hat{\sigma}_v$	Functions
A_{1g}	1	1	1	1	1	1	1	1	1	1	1	1	$x^2 + y^2, z^2$
A_{2g}	1	1	1	1	-1	-1	1	1	1	1	-1	-1	
B_{1g}	1	-1	1	-1	1	-1	1	-1	1	-1	1	-1	
B_{2g}	1	-1	1	-1	-1	1	1	-1	1	-1	-1	1	
E_{1g}	2	1	-1	-2	0	0	2	1	-1	-2	0	0	(xz, yz)
E_{2g}	2	-1	-1	2	0	0	2	-1	-1	2	0	0	$(x^2 - y^2, xy)$
A_{1u}	1	1	1	1	1	1	-1	-1	-1	-1	-1	-1	
A_{2u}	1	1	1	1	-1	-1	-1	-1	-1	-1	1	1	z
B_{1u}	1	-1	1	-1	1	-1	-1	1	-1	1	-1	1	
B_{2u}	1	-1	1	-1	-1	1	-1	1	-1	1	1	-1	
E_{1u}	2	1	-1	-2	0	0	-2	-1	1	2	0	0	(x, y)
E_{2u}	2	-1	-1	2	0	0	-2	1	1	-2	0	0	

O_h	$\hat{E}$	$8\hat{C}_3$	$6\hat{C}_2$	$6\hat{C}_4$	$3\hat{C}_2$	$\hat{i}$	$6\hat{S}_4$	$8\hat{S}_6$	$3\hat{\sigma}_h$	$6\hat{\sigma}_d$	Functions
A_{1g}	1	1	1	1	1	1	1	1	1	1	$x^2 + y^2, z^2$
A_{2g}	1	1	-1	-1	1	1	-1	1	1	-1	
E_g	2	-1	0	0	2	2	0	-1	2	0	$(3z^2 - r^2, x^2 - y^2)$
T_{1g}	3	0	-1	1	-1	3	1	0	-1	-1	
T_{2g}	3	0	1	-1	-1	3	-1	0	-1	1	(xz, yz, xy)
A_{1u}	1	1	1	1	1	-1	-1	-1	-1	-1	
A_{2u}	1	1	-1	-1	1	-1	1	-1	-1	1	
E_u	2	-1	0	0	2	-2	0	1	-2	0	
T_{1u}	3	0	-1	1	-1	-3	-1	0	1	1	(x, y, z)
T_{2u}	3	0	1	-1	-1	-3	1	0	1	-1	

Table 14.4 Average Bond Energies and Dipoles

Bond	Energy kJ/mol	Dipole debye	Bond	Energy kJ/mol	Dipole debye
C—C	344	0	H—H	435	0
C—O	350	0.8	H—N	390	1.3
C—N	293	0.5	H—O	464	1.5
C=C	615	0	H—C	415	0.4
C=O	724	2.5	C≡N	891	3.5
C=N	615	—	N≡N	942	0
C≡C	812	0	O=O	492	0
C—Cl	326	1.5	Cl—Cl	239	0

Table 14.5 Some Common Hybrid Orbitals

Coordination Number	Hybrid Description for Central Atom	Geometry	Bond Angles
2	sp	Linear	180°
3	sp^2	Trigonal planar	120°
4	sp^3	Tetrahedral	109°28'
4	dsp^2	Square planar	90°
5	dsp^3	Trigonal bipyramid	120°, 90°
6	d^2sp^3	Octahedral	90°

Table 14.9 States of Equivalent Electrons for Open-Shell Degenerate Configurations

	Configuration	States
$C_{\infty v}(D_{\infty h})^a$	π^2	$^1\Sigma^+, {}^1\Delta, {}^3\Sigma^-$
C_{3v}	e^2	$^1A_1, {}^1E, {}^3A_2$
D_{6h}	e_{1g}^2 or e_{1u}^2	$^1A_{1g}, {}^1E_{2g}, {}^3A_{2g}$
	e_{2g}^2 or e_{2u}^2	$^1A_{1g}, {}^1E_{2g}, {}^3A_{2g}$
$T_d(O)_h{}^a$	e^2	$^1A_1, {}^1E, {}^3A_2$
	t_1^2, t_1^4	$^1A_1, {}^1E, {}^1T_2, {}^3T_1$
	t_1^3	$^2E, {}^2T_1, {}^2T_2, {}^4A_1$
	t_2^4, t_2^2	$^1A_1, {}^1E, {}^1T_2, {}^3T_1$
	t_2^3	$^2E, {}^2T_1, {}^2T_2 {}^4A_2$

^{a} For $D_{\infty h}$ and O_h, add g/u to symbols as appropriate; for example O_h, e_u^2 gives the states $^1A_{1g}, {}^1E_g, {}^3A_{2g}$.

Table 14.10 Vibrational Constants of Some Molecules

	$\omega_{e1}/\text{cm}^{-1}$	$\omega_{e2}/\text{cm}^{-1}$	$\omega_{e3}/\text{cm}^{-1}$	$\omega_{e4}/\text{cm}^{-1}$
Triatomic C_{2v}	a_1	a_1	b_1	
H$_2$O	3652	1595	3756	
D$_2$O	2666	1179	2784	
H$_2$S	2611	1290	2684	
CH$_2$	2968	1444	3000	
SO$_2$	1151	524	1361	
F$_2$O	830	490	1110	
Linear $C_{\infty v}$	σ^+	π	σ^+	
HCN	2089	712	3312	
ClCN	729	397	2201	
OCS	859	527	2079	
NNO	1285	589	2224	
Linear $D_{\infty h}$	σ_g^+	π_u	σ_u^+	
CO$_2$	1388	667	2349	
CS$_2$ 657	397	1523		
Pyramidal C_{3v}	a_1	a_1	e	e
NH$_3$	3337	950	3414	1628
NO$_3$	2419	749	2555	1191
PH$_3$	2327	991	2421	1121
Tetrahedral T_d	a_1	e	t_2	t_2
CH$_4$	2914	1526	3020	1306
CD$_4$	2085	1054	2258	996
SiH$_4$	2187	978	2183	910
SiF$_4$	800	260	1022	420

$$I = \sum_i m_i r_i^2 \qquad \textbf{(14.51)} \qquad \hat{H} = \frac{\hbar^2}{2}\left[\frac{\hat{J}_X^2}{I_{XX}} + \frac{\hat{J}_Y^2}{I_{YY}} + \frac{\hat{J}_Z^2}{I_{ZZ}}\right] \qquad \textbf{(14.52)}$$

Table 14.12   Rotational Constants*

Linear	B_e/GHz	Symmetric Tops	B/GHz
HCN	44.31597	$^{14}NF_3$	10.68107
DCN	36.20740	$^{15}NF_3$	10.62935
$^{79}BrCN$	4.120198	PF_3	7.82001
$^{81}BrCN$	4.096788	$P^{35}Cl_3$	2.6171
OCS	6.081494	CH_3F	25.53591
$OC^{34}S$	5.93284	$CH_3{}^{35}Cl$	13.29295
$OC^{36}S$	5.79967	$CH_3{}^{37}Cl$	13.08824
^{18}OCS	5.70483		
$O^{13}CS$	5.69095		

Asymmetric Tops	A/GHz	B/GHz	C/GHz
HDS	290.300	145.200	94.130
O_3	106.530	13.349	11.843
CH_2O	282.106	38.834	34.004
CH_2F_2	49.138	10.604	9.249

* If no mass number is given, the most common isotope is intended; for example: ^{16}O, ^{32}S, ^1H, ^{14}N, ^{12}C.

CHAPTER 15

distinguishable
$$Z = z^N \qquad \textbf{(15.18b)}$$

fermions
$$n_i = \frac{Ae^{-\epsilon_i/k_bT}}{1 + Ae^{-\epsilon_i/k_bT}} \qquad \textbf{(15.19a)}$$

bosons
$$n_i = \frac{Ae^{-\epsilon_i/k_bT}}{1 - Ae^{-\epsilon_i/k_bT}} \qquad \textbf{(15.19b)}$$

indistinguishable
$$Z = \frac{z^N}{N!} \qquad \textbf{(15.20)}$$

Table 15.1   Nuclear Spin Quantum Numbers

Nucleus	I	$g_I = 2I + 1$
^{12}C, ^{16}O, ^{32}S	0	1
^1H, ^{13}C, ^{19}F, ^{31}P, ^{195}Pt, ^{207}Pb	$\frac{1}{2}$	2
^2D, ^{14}N, ^6Li	1	3
^7Li, ^{11}B, ^{23}Na, ^{35}Cl, ^{37}Cl	$\frac{3}{2}$	4
^{27}Al, ^{55}Mn	$\frac{5}{2}$	6
^{10}B	3	7
^{115}In, ^{45}Sc	$\frac{7}{2}$	8

Einstein
$$C_{vm} = \left(\frac{\partial U_m}{\partial T}\right)_V$$

$$C_{vm} = \frac{3Ru^2e^u}{(e^u - 1)^2} \tag{15.46}$$

$$\theta_E = \frac{h\nu_E}{k_b}; \qquad u = \frac{\theta_E}{T} \tag{15.47}$$

Debye
$$C_{vm} = 3RD(\theta_D/T) \tag{15.57}$$

$$D(\theta_D/T) = 3\left(\frac{T}{\theta_D}\right)^3 \int_0^{\theta_D/T} \frac{u^4e^u\,du}{(e^u - 1)^2} \tag{15.58}$$

Debye–Sommerfeld
$$C_{vm} = \frac{36\pi^4 R}{15\theta_D^3} T^3 - \gamma T \tag{15.60}$$

Table 15.4 Heat-Capacity Constants for Solids

Metals*			Nonmetals	
θ_D/K	$(10^5\gamma R)/\text{K}^{-1}$			θ_D/K
Ag 225	7.33		Diamond	1860
Al 426	16.4		KCl	227
Be 1160	2.7		NaCl	281
Na 158	17.		AgCl	183
Ni 440	84.7		AgBr	144
Pt 240	80.0		CaF$_2$	474
Pb 96	37.6		FeS$_2$	645
Zn 300	7.5			

*Source: *Handbook of Chemistry and Physics,* 48 ed., 1967: The Chemical Rubber Co., Cleveland, Ohio.

$$(T > \theta_D) \qquad C_{vm} = 3R\left(1 - \frac{\theta_D^2}{20T^2}\right) \tag{15.61}$$

See Figure 15.10 on page 55.

Planck $u = \dfrac{h\nu}{k_b T}$
$$\rho(\nu)\,d\nu = \frac{8\pi k_b^4 T^4}{h^3 c^3} \frac{u^3\,du}{e^u - 1} \tag{15.64}$$

Stefan–Boltzmann
$$e(T) = \left(\frac{2\pi^5 k_b^4}{15h^3c^2}\right) T^4 \tag{15.66}$$

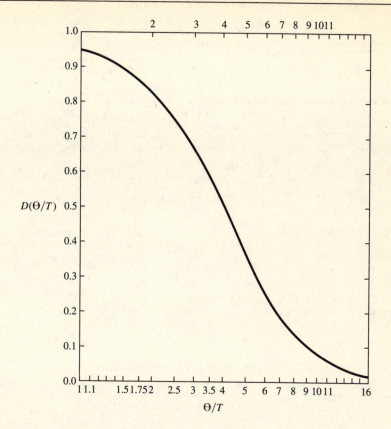

Figure 15.10 The Debye heat-capacity function. This graph can be used to calculate heat capacities of atomic solids for which the Debye characteristic temperature (θ) is known. Similarly, a single measurement of such a heat capacity (provided it is made in the temperature range where the heat capacity is changing significantly) can be used to calculate the Debye function (D, vertical scale) using Eq. (15.57); then this graph will give the value of the characteristic temperature and, hence, the heat capacity at other temperatures.

CHAPTER 16

Table 16.1 Crystal Systems

System	Lattice Constants	Angle Restrictions
Cubic*	$a = b = c$	$\alpha = \beta = \gamma = 90°$
Hexagonal	$a = b \neq c$	$\alpha = \beta = 90°;\ \gamma = 120°$
Rhombohedral	$a = b = c$	$\alpha = \beta = \gamma$
Tetragonal	$a = b \neq c$	$\alpha = \beta = \gamma = 90°$
Orthorhombic	$a \neq b \neq c$	$\alpha = \beta = \gamma = 90°$
Monoclinic	$a \neq b \neq c$	$\alpha = \gamma = 90°$
Triclinic	$a \neq b \neq c$	$\alpha \neq \beta \neq \gamma$

*Also known as *isometric*.

$$\cos \theta_a = \cos \theta_{a_0} - h\lambda/a$$
$$\cos \theta_b = \cos \theta_{b_0} - k\lambda/b$$
$$\cos \theta_c = \cos \theta_{c_0} - l\lambda/c \qquad \textbf{(16.4)}$$

$$2d \sin \theta = n\lambda \qquad \textbf{(16.6)} \qquad\qquad d_{hkl} \sin \theta = \lambda/2 \qquad \textbf{(16.7)}$$

Table 16.3   Ionic Crystal Radii (unit: pm—divide by 100 for Å)

Li^+	60	Be^{2+}	31					B^{3+} 20	C^{4+} 15	O^{2-} 140	F^- 136
Na^+	95	Mg^{2+}	65					Al^{3+} 50	Si^{4+} 41	S^{2-} 184	Cl^- 181
K^+	133	Ca^{2+}	99	Cu^+	96	Zn^{2+}	74			Se^{2-} 198	Br^- 195
Rb^+	148	Sr^{2+}	113	Ag^+	126	Cd^{2+}	97				I^- 216
Cs^+	169	Ba^{2+}	135	Au^+	137	Hg^{2+}	216				
		Fe^{2+}	74	Cu^{2+}	72	Fe^{3+}	64				

Table 16.4   Madelung Constants

AB types:	Cesium chloride	1.76267
	Rock salt (NaCl)	1.74756
	Zinc blende (ZnS)	1.63805
	Zincite (ZnO)	1.64132
AB$_2$ types:	Fluorite (CaCl$_2$)	5.03878
	Rutile (TiO$_2$)	4.7701

Table 16.5   Enthalpies of Sublimation of Alkali Metals (kJ/mol)

Li	161.5	Rb	85.8
Na	110.2	Cs	78.7
K	90.0		

Table 16.6   Grant–Paul Parameters for C-13 Chemical Shifts*

Position	Shift	Connection	Shift	Connection	Shift
α	8.85	1°(3°)	−0.96	4°(1°)	−1.27
β	9.51	1°(4°)	−3.61	4°(2°)	−8.24
γ	−2.34	2°(3°)	−2.11		
δ	0.28	2°(4°)	−7.14		
ϵ	0.03	3°(2°)	−3.04	3°(3°)	−9.05
constant	−2.35				

*C. J. Carman, *Macromolecules*, **6**, 719 (1973). Shifts in ppm relative to internal TMS.

APPENDIX I

$$e^x = 1 + x + \frac{x^2}{2!} + \frac{x^3}{3!} + \frac{x^4}{4!} + \cdots \qquad \textbf{(AI.3)}$$

$$\ln(1 + x) = x - \frac{x^2}{2} + \frac{x^3}{3} - \frac{x^4}{4} + \cdots \qquad (|x| \le 1) \quad \textbf{(AI.4)}$$

$$\frac{1}{1 + x} = 1 - x + x^2 - x^3 + x^4 - \cdots \qquad (|x| < 1) \quad \textbf{(AI.5)}$$

$$(1 + x)^{1/2} = 1 + \frac{x}{2} - \frac{x^2}{8} + \frac{x^3}{10} - \frac{5x^4}{128} + \frac{7x^5}{256} - \frac{21x^6}{1024} + \cdots \quad (|x| < 1) \quad \textbf{(AI.6)}$$

$$x = \frac{-B \pm \sqrt{B^2 - 4AC}}{2A} \tag{AI.7}$$

$$q = -\frac{1}{2}[B + \text{sgn}(B)\sqrt{B^2 - 4AC}] \quad \text{roots:} \quad x = \frac{q}{A} \quad \text{and} \quad \frac{C}{q} \tag{AI.8}$$

Newton–Raphson
$$x_{n+1} = x_n - \frac{f(x_n)}{f'(x_n)} \tag{AI.9}$$

$$\int_{x_0}^{x_1} f(x) \, dx \cong \tfrac{1}{2}(f_1 + f_0)(x_1 - x_0) \tag{AI.10}$$

$$\int_{x_0}^{x_n} f(x) \, dx = \frac{h}{3}[f_0 + 4f_1 + 2f_2 + 4f_3 + 2f_4 + \cdots + 4f_{n-1} + f_n] \tag{AI.11}$$

$$\frac{df}{dx} = \frac{f(x + h) - f(x - h)}{2h} \tag{AI.14}$$

$$m = \frac{N\sum (x_i y_i) - \left(\sum x_i\right)\left(\sum y_i\right)}{D}$$

$$a = \frac{\left(\sum y_i\right)\left(\sum x_i^2\right) - \sum (x_i y_i)\left(\sum x_i\right)}{D} \tag{AI.20}$$

$$D = N\left(\sum x_i^2\right) - \left(\sum x_i\right)^2$$

$$r = \frac{m\sigma_x}{\sigma_y} \tag{AI.22}$$

$$\sigma_{\text{fit}} = \sqrt{\frac{R}{DF}} = \left[\frac{\sum (y_i - y_{\text{calc}})^2}{N - 2}\right]^{1/2} \tag{AI.23}$$

$$\sigma_m = \frac{m}{r}\sqrt{\frac{1 - r^2}{N - 2}} \tag{AI.25}$$

$$\sigma_a = \sigma_m\sqrt{\frac{\sum x_i^2}{N}} \tag{AI.26}$$

$$\lambda = t_c \sigma \tag{AI.27}$$

Table AI.1   Critical *t* Factors

Degrees of Freedom	$t_c(90\%)$	$t_c(95\%)$	$t_c(99\%)$
1	6.31	12.7	63.7
2	2.92	4.30	9.92
3	2.35	3.18	5.84
4	2.13	2.78	4.60
5	2.01	2.57	4.03
6	1.94	2.45	3.71
8	1.86	2.31	3.36
10	1.81	2.23	3.17
15	1.75	2.13	2.95
20	1.72	2.09	2.85
30	1.70	2.04	2.75
∞	1.64	1.96	2.58

APPENDIX II

$$dh = \left(\frac{\partial h}{\partial x}\right)_y dx + \left(\frac{\partial h}{\partial y}\right)_x dy \qquad \textbf{(AII.5a)}$$

CHAIN RULE
$$\left(\frac{\partial z}{\partial w}\right)_y = \left(\frac{\partial z}{\partial x}\right)_y\left(\frac{\partial x}{\partial w}\right)_y \qquad \textbf{(AII.7)}$$

RECIPROCAL RULE
$$\left(\frac{\partial z}{\partial x}\right)_y \left(\frac{\partial x}{\partial z}\right)_y = 1 \qquad \textbf{(AII.9)}$$

CYCLIC RULE
$$\left(\frac{\partial x}{\partial y}\right)_z = -\left(\frac{\partial x}{\partial z}\right)_y \left(\frac{\partial z}{\partial y}\right)_x \qquad \textbf{(AII.10)}$$

$$\left(\frac{\partial z}{\partial x}\right)_w = \left(\frac{\partial z}{\partial x}\right)_y + \left(\frac{\partial z}{\partial y}\right)_x \left(\frac{\partial y}{\partial x}\right)_w \qquad \textbf{(AII.11)}$$

APPENDIX III

$$x = r \sin\theta \cos\phi$$
$$y = r \sin\theta \sin\phi$$
$$z = r \cos\theta \qquad \textbf{(AIII.5)}$$
$$d\tau = r^2\,dr \sin\theta\,d\theta\,d\phi \qquad \textbf{(AIII.7)}$$
$$d\Omega = \sin\theta\,d\theta\,d\phi \qquad \textbf{(AIII.9)}$$

APPENDIX IV ▓▓▓

$$e^{i\phi} = \cos \phi + i \sin \phi \qquad \textbf{(AIV.17)}$$

$$e^{-i\phi} = \cos \phi - i \sin \phi \qquad \textbf{(AIV.18)}$$

$$\cos \phi = \frac{1}{2}(e^{i\phi} + e^{-i\phi}) \qquad \textbf{(AIV.19)}$$

$$\sin \phi = \frac{1}{2i}(e^{i\phi} - e^{-i\phi}) \qquad \textbf{(AIV.20)}$$

APPENDIX V ▓▓

$$P_{ij} = \sum_{k} A_{ik} B_{kj} \qquad \textbf{(AV.1)}$$